FREE Test Taking Tips DVD Offer

To help us better serve you, we have developed a Test Taking Tips DVD that we would like to give you for FREE. **This DVD covers world-class test taking tips that you can use to be even more successful when you are taking your test.**

All that we ask is that you email us your feedback about your study guide. Please let us know what you thought about it – whether that is good, bad or indifferent.

To get your **FREE Test Taking Tips DVD**, email freedvd@studyguideteam.com with "FREE DVD" in the subject line and the following information in the body of the email:

a. The title of your study guide.

b. Your product rating on a scale of 1-5, with 5 being the highest rating.

c. Your feedback about the study guide. What did you think of it?

d. Your full name and shipping address to send your free DVD.

If you have any questions or concerns, please don't hesitate to contact us at freedvd@studyguideteam.com.

Thanks again!

Praxis 5169 Middle School Math Study Guide

Praxis II Middle School Mathematics 5169 Exam Prep and Practice Test Questions
[2nd Edition]

TPB Publishing

Written and edited by Test Prep Books.

ISBN 13: 9781628458497
ISBN 10: 1628458496

Table of Contents

Quick Overview

As you draw closer to taking your exam, effective preparation becomes more and more important. Thankfully, you have this study guide to help you get ready. Use this guide to help keep your studying on track and refer to it often.

This study guide contains several key sections that will help you be successful on your exam. The guide contains tips for what you should do the night before and the day of the test. Also included are test-taking tips. Knowing the right information is not always enough. Many well-prepared test takers struggle with exams. These tips will help equip you to accurately read, assess, and answer test questions.

A large part of the guide is devoted to showing you what content to expect on the exam and to helping you better understand that content. In this guide are practice test questions so that you can see how well you have grasped the content. Then, answer explanations are provided so that you can understand why you missed certain questions.

Don't try to cram the night before you take your exam. This is not a wise strategy for a few reasons. First, your retention of the information will be low. Your time would be better used by reviewing information you already know rather than trying to learn a lot of new information. Second, you will likely become stressed as you try to gain a large amount of knowledge in a short amount of time. Third, you will be depriving yourself of sleep. So be sure to go to bed at a reasonable time the night before. Being well-rested helps you focus and remain calm.

Be sure to eat a substantial breakfast the morning of the exam. If you are taking the exam in the afternoon, be sure to have a good lunch as well. Being hungry is distracting and can make it difficult to focus. You have hopefully spent lots of time preparing for the exam. Don't let an empty stomach get in the way of success!

When travelling to the testing center, leave earlier than needed. That way, you have a buffer in case you experience any delays. This will help you remain calm and will keep you from missing your appointment time at the testing center.

Be sure to pace yourself during the exam. Don't try to rush through the exam. There is no need to risk performing poorly on the exam just so you can leave the testing center early. Allow yourself to use all of the allotted time if needed.

Remain positive while taking the exam even if you feel like you are performing poorly. Thinking about the content you should have mastered will not help you perform better on the exam.

Once the exam is complete, take some time to relax. Even if you feel that you need to take the exam again, you will be well served by some down time before you begin studying again. It's often easier to convince yourself to study if you know that it will come with a reward!

Test-Taking Strategies

1. Predicting the Answer

When you feel confident in your preparation for a multiple-choice test, try predicting the answer before reading the answer choices. This is especially useful on questions that test objective factual knowledge. By predicting the answer before reading the available choices, you eliminate the possibility that you will be distracted or led astray by an incorrect answer choice. You will feel more confident in your selection if you read the question, predict the answer, and then find your prediction among the answer choices. After using this strategy, be sure to still read all of the answer choices carefully and completely. If you feel unprepared, you should not attempt to predict the answers. This would be a waste of time and an opportunity for your mind to wander in the wrong direction.

2. Reading the Whole Question

Too often, test takers scan a multiple-choice question, recognize a few familiar words, and immediately jump to the answer choices. Test authors are aware of this common impatience, and they will sometimes prey upon it. For instance, a test author might subtly turn the question into a negative, or he or she might redirect the focus of the question right at the end. The only way to avoid falling into these traps is to read the entirety of the question carefully before reading the answer choices.

3. Looking for Wrong Answers

Long and complicated multiple-choice questions can be intimidating. One way to simplify a difficult multiple-choice question is to eliminate all of the answer choices that are clearly wrong. In most sets of answers, there will be at least one selection that can be dismissed right away. If the test is administered on paper, the test taker could draw a line through it to indicate that it may be ignored; otherwise, the test taker will have to perform this operation mentally or on scratch paper. In either case, once the obviously incorrect answers have been eliminated, the remaining choices may be considered. Sometimes identifying the clearly wrong answers will give the test taker some information about the correct answer. For instance, if one of the remaining answer choices is a direct opposite of one of the eliminated answer choices, it may well be the correct answer. The opposite of obviously wrong is obviously right! Of course, this is not always the case. Some answers are obviously incorrect simply because they are irrelevant to the question being asked. Still, identifying and eliminating some incorrect answer choices is a good way to simplify a multiple-choice question.

4. Don't Overanalyze

Anxious test takers often overanalyze questions. When you are nervous, your brain will often run wild, causing you to make associations and discover clues that don't actually exist. If you feel that this may be a problem for you, do whatever you can to slow down during the test. Try taking a deep breath or counting to ten. As you read and consider the question, restrict yourself to the particular words used by the author. Avoid thought tangents about what the author *really* meant, or what he or she was *trying* to say. The only things that matter on a multiple-choice test are the words that are actually in the question. You must avoid reading too much into a multiple-choice question, or supposing that the writer meant something other than what he or she wrote.

5. No Need for Panic

It is wise to learn as many strategies as possible before taking a multiple-choice test, but it is likely that you will come across a few questions for which you simply don't know the answer. In this situation, avoid panicking. Because most multiple-choice tests include dozens of questions, the relative value of a single wrong answer is small. As much as possible, you should compartmentalize each question on a multiple-choice test. In other words, you should not allow your feelings about one question to affect your success on the others. When you find a question that you either don't understand or don't know how to answer, just take a deep breath and do your best. Read the entire question slowly and carefully. Try rephrasing the question a couple of different ways. Then, read all of the answer choices carefully. After eliminating obviously wrong answers, make a selection and move on to the next question.

6. Confusing Answer Choices

When working on a difficult multiple-choice question, there may be a tendency to focus on the answer choices that are the easiest to understand. Many people, whether consciously or not, gravitate to the answer choices that require the least concentration, knowledge, and memory. This is a mistake. When you come across an answer choice that is confusing, you should give it extra attention. A question might be confusing because you do not know the subject matter to which it refers. If this is the case, don't eliminate the answer before you have affirmatively settled on another. When you come across an answer choice of this type, set it aside as you look at the remaining choices. If you can confidently assert that one of the other choices is correct, you can leave the confusing answer aside. Otherwise, you will need to take a moment to try to better understand the confusing answer choice. Rephrasing is one way to tease out the sense of a confusing answer choice.

7. Your First Instinct

Many people struggle with multiple-choice tests because they overthink the questions. If you have studied sufficiently for the test, you should be prepared to trust your first instinct once you have carefully and completely read the question and all of the answer choices. There is a great deal of research suggesting that the mind can come to the correct conclusion very quickly once it has obtained all of the relevant information. At times, it may seem to you as if your intuition is working faster even than your reasoning mind. This may in fact be true. The knowledge you obtain while studying may be retrieved from your subconscious before you have a chance to work out the associations that support it. Verify your instinct by working out the reasons that it should be trusted.

8. Key Words

Many test takers struggle with multiple-choice questions because they have poor reading comprehension skills. Quickly reading and understanding a multiple-choice question requires a mixture of skill and experience. To help with this, try jotting down a few key words and phrases on a piece of scrap paper. Doing this concentrates the process of reading and forces the mind to weigh the relative importance of the question's parts. In selecting words and phrases to write down, the test taker thinks about the question more deeply and carefully. This is especially true for multiple-choice questions that are preceded by a long prompt.

9. Subtle Negatives

One of the oldest tricks in the multiple-choice test writer's book is to subtly reverse the meaning of a question with a word like *not* or *except*. If you are not paying attention to each word in the question, you can easily be led astray by this trick. For instance, a common question format is, "Which of the following is…?" Obviously, if the question instead is, "Which of the following is not…?," then the answer will be quite different. Even worse, the test makers are aware of the potential for this mistake and will include one answer choice that would be correct if the question were not negated or reversed. A test taker who misses the reversal will find what he or she believes to be a correct answer and will be so confident that he or she will fail to reread the question and discover the original error. The only way to avoid this is to practice a wide variety of multiple-choice questions and to pay close attention to each and every word.

10. Reading Every Answer Choice

It may seem obvious, but you should always read every one of the answer choices! Too many test takers fall into the habit of scanning the question and assuming that they understand the question because they recognize a few key words. From there, they pick the first answer choice that answers the question they believe they have read. Test takers who read all of the answer choices might discover that one of the latter answer choices is actually *more* correct. Moreover, reading all of the answer choices can remind you of facts related to the question that can help you arrive at the correct answer. Sometimes, a misstatement or incorrect detail in one of the latter answer choices will trigger your memory of the subject and will enable you to find the right answer. Failing to read all of the answer choices is like not reading all of the items on a restaurant menu: you might miss out on the perfect choice.

11. Spot the Hedges

One of the keys to success on multiple-choice tests is paying close attention to every word. This is never truer than with words like almost, most, some, and sometimes. These words are called "hedges" because they indicate that a statement is not totally true or not true in every place and time. An absolute statement will contain no hedges, but in many subjects, the answers are not always straightforward or absolute. There are always exceptions to the rules in these subjects. For this reason, you should favor those multiple-choice questions that contain hedging language. The presence of qualifying words indicates that the author is taking special care with his or her words, which is certainly important when composing the right answer. After all, there are many ways to be wrong, but there is only one way to be right! For this reason, it is wise to avoid answers that are absolute when taking a multiple-choice test. An absolute answer is one that says things are either all one way or all another. They often include words like *every*, *always*, *best*, and *never*. If you are taking a multiple-choice test in a subject that doesn't lend itself to absolute answers, be on your guard if you see any of these words.

12. Long Answers

In many subject areas, the answers are not simple. As already mentioned, the right answer often requires hedges. Another common feature of the answers to a complex or subjective question are qualifying clauses, which are groups of words that subtly modify the meaning of the sentence. If the question or answer choice describes a rule to which there are exceptions or the subject matter is complicated, ambiguous, or confusing, the correct answer will require many words in order to be expressed clearly and accurately. In essence, you should not be deterred by answer choices that seem excessively long. Oftentimes, the author of the text will not be able to write the correct answer without offering some qualifications and modifications. Your job is to read the answer choices thoroughly and completely and to select the one that most accurately and precisely answers the question.

13. Restating to Understand

Sometimes, a question on a multiple-choice test is difficult not because of what it asks but because of how it is written. If this is the case, restate the question or answer choice in different words. This process serves a couple of important purposes. First, it forces you to concentrate on the core of the question. In order to rephrase the question accurately, you have to understand it well. Rephrasing the question will concentrate your mind on the key words and ideas. Second, it will present the information to your mind in a fresh way. This process may trigger your memory and render some useful scrap of information picked up while studying.

14. True Statements

Sometimes an answer choice will be true in itself, but it does not answer the question. This is one of the main reasons why it is essential to read the question carefully and completely before proceeding to the answer choices. Too often, test takers skip ahead to the answer choices and look for true statements. Having found one of these, they are content to select it without reference to the question above. Obviously, this provides an easy way for test makers to play tricks. The savvy test taker will always read the entire question before turning to the answer choices. Then, having settled on a correct answer choice, he or she will refer to the original question and ensure that the selected answer is relevant. The mistake of choosing a correct-but-irrelevant answer choice is especially common on questions related to specific pieces of objective knowledge. A prepared test taker will have a wealth of factual knowledge at his or her disposal, and should not be careless in its application.

15. No Patterns

One of the more dangerous ideas that circulates about multiple-choice tests is that the correct answers tend to fall into patterns. These erroneous ideas range from a belief that B and C are the most common right answers, to the idea that an unprepared test-taker should answer "A-B-A-C-A-D-A-B-A." It cannot be emphasized enough that pattern-seeking of this type is exactly the WRONG way to approach a multiple-choice test. To begin with, it is highly unlikely that the test maker will plot the correct answers according to some predetermined pattern. The questions are scrambled and delivered in a random order. Furthermore, even if the test maker was following a pattern in the assignation of correct answers, there is no reason why the test taker would know which pattern he or she was using. Any attempt to discern a pattern in the answer choices is a waste of time and a distraction from the real work of taking the test. A test taker would be much better served by extra preparation before the test than by reliance on a pattern in the answers.

FREE DVD OFFER

Don't forget that doing well on your exam includes both understanding the test content and understanding how to use what you know to do well on the test. We offer a completely FREE Test Taking Tips DVD that covers world class test taking tips that you can use to be even more successful when you are taking your test.

All that we ask is that you email us your feedback about your study guide. To get your **FREE Test Taking Tips DVD**, email freedvd@studyguideteam.com with "FREE DVD" in the subject line and the following information in the body of the email:

- The title of your study guide.
- Your product rating on a scale of 1-5, with 5 being the highest rating.
- Your feedback about the study guide. What did you think of it?
- Your full name and shipping address to send your free DVD.

Introduction to Praxis Middle School Mathematics

Function of the Praxis Middle School Math Test

The Praxis Middle School Mathematics (5169) test is one of the Educational Testing Service's (ETS's) Subject Assessment tests that measures specific teaching skills and knowledge. It is intended to be used as part of the certification process for teachers of mathematics at the middle school level. Most test takers have a bachelor's degree in mathematics, education, or mathematics education specifically. Recommended course work prior to sitting for the exam should include courses in the major domains of mathematics (arithmetic and mathematics foundations, algebra and functions, geometry and measurements, and probability and statistics) for elementary and/or middle school teachers. An overview of calculus, trigonometry, and logic is helpful as well.

The Praxis Middle School Mathematics test requires test takers to understand and apply mathematical concepts, identify and extend patterns, reason mathematically and make conjectures, develop solvable models to describe real-world scenarios, and solve problems that may require integrating knowledge from different domains of mathematics.

The requirements to obtain certification as a middle school math teacher vary by state, but the Praxis Middle School Mathematics test is designed to be consistent with the recommendations for national standards for mathematics education.

Test Administration

All Praxis Subject Assessment Tests, including the Middle School Math test, are only administered in electronic format and in the English language. However, a version written in Braille is provided for those with certain visual impairments. Although there are not testing administration accommodations for other languages, extended testing time may be granted for candidates for whom English is a secondary language.

Depending upon the state requirements, tests may be offered continuously on a rolling basis throughout the year, or only administered during certain windows during the year. Prior to sitting for the exam, candidates must first register for the test online, submit the applicable fees, then report to the testing center.

Individuals with disabilities who would like test accommodations should register through the ETS Disability Services to have the requested accommodations approved. Online registration is not available for disability accommodations; applicants must make requests via mail or email. It takes approximately six weeks for the documentation to be reviewed. Accommodation request procedures are outlined in the *Bulletin Supplement for Test Takers with Disabilities or Health-related Needs.* Although online registration is not available, test scores may be viewed online, for which an account must be created.

Test Format

The Praxis Middle School Math test is comprised of 55 questions total; 34 of them pertain to Arithmetic and Algebra, and 21 concern Geometry and Measurement topics. Questions are of multiple-choice and numeric entry answer formats. Test takers are allotted two hours to complete the exam and are provided with an on-screen graphing calculator.

Scores

The Praxis Middle School Mathematics (5169) test, which replaced the Middle School Mathematics (0069) test in 2013, is scored on a scale from 100-200 points. Scores are calculated based on the number of correct responses. The median score in the past several years for the exam has been approximately 165. The minimum score required to pass the exam depends on the state in which the test taker is seeking certification because each state can set their own passing score. Score reports will indicate the test taker's passing status for the states he or she indicates on the exam registration form.

Arithmetic and Algebra

Operations and Properties of the Real Number System

Basic Operations of Arithmetic

There are four different basic operations used with numbers: addition, subtraction, multiplication, and division.

- Addition takes two numbers and combines them into a total called the sum. The sum is the total when combining two collections into one. If there are 5 things in one collection and 3 in another, then after combining them, there is a total of $5 + 3 = 8$. Note the order does not matter when adding numbers. For example, $3 + 5 = 8$.

- Subtraction is the opposite (or "inverse") operation to addition. Whereas addition combines two quantities together, subtraction takes one quantity away from another. For example, if there are 20 gallons of fuel and 5 are removed, that gives $20 - 5 = 15$ gallons remaining. Note that for subtraction, the order does matter because it makes a difference which quantity is being removed from which.

- Multiplication is repeated addition. 3×4 can be thought of as putting together 3 sets of items, each set containing 4 items. The total is 12 items. Another way to think of this is to think of each number as the length of one side of a rectangle. If a rectangle is covered in tiles with 3 columns of 4 tiles each, then there are 12 tiles in total. From this, one can see that the answer is the same if the rectangle has 4 rows of 3 tiles each: $4 \times 3 = 12$. By expanding this reasoning, the order the numbers are multiplied does not matter.

- Division is the opposite of multiplication. It means taking one quantity and dividing it into sets the size of the second quantity. If there are 16 sandwiches to be distributed to 4 people, then each person gets $16 \div 4 = 4$ sandwiches. As with subtraction, the order in which the numbers appear does matter for division.

<u>Addition</u>
Addition is the combination of two numbers so their quantities are added together cumulatively. The sign for an addition operation is the + symbol. For example, 9 + 6 = 15. The 9 and 6 combine to achieve a cumulative value, called a **sum**.

Addition holds the commutative property, which means that the order of the numbers in an addition equation can be switched without altering the result. The formula for the commutative property is a + b = b + a. Let's look at a few examples to see how the commutative property works:

$$7 = 3 + 4 = 4 + 3 = 7$$

$$20 = 12 + 8 = 8 + 12 = 20$$

Addition also holds the **associative property**, which means that the grouping of numbers doesn't matter in an addition problem. In other words, the presence or absence of parentheses is irrelevant. The formula for the associative property is $(a + b) + c = a + (b + c)$. Here are some examples of the associative property at work:

$$30 = (6 + 14) + 10 = 6 + (14 + 10) = 30$$

$$35 = 8 + (2 + 25) = (8 + 2) + 25 = 35$$

Subtraction

Subtraction is taking away one number from another, so their quantities are reduced. The sign designating a subtraction operation is the – symbol, and the result is called the **difference**. For example, $9 - 6 = 3$. The number *6* detracts from the number *9* to reach the difference *3*.

Unlike addition, subtraction follows neither the commutative nor associative properties. The order and grouping in subtraction impact the result.

$$15 = 22 - 7 \neq 7 - 22 = -15$$

$$3 = (10 - 5) - 2 \neq 10 - (5 - 2) = 7$$

When working through subtraction problems involving larger numbers, it's necessary to regroup the numbers. Let's work through a practice problem using regrouping:

$$\begin{array}{r} 3\ 2\ 5 \\ -\ \ 7\ 7 \\ \hline \end{array}$$

Here, it is clear that the ones and tens columns for 77 are greater than the ones and tens columns for 325. To subtract this number, borrow from the tens and hundreds columns. When borrowing from a column, subtracting 1 from the lender column will add 10 to the borrower column:

$$\begin{array}{rrr} 3-1 & 10+2-1 & 10+5 \\ - & 7 & 7 \\ \hline \end{array} = \begin{array}{rrr} 2 & 11 & 15 \\ - & 7 & 7 \\ \hline 2 & 4 & 8 \end{array}$$

After ensuring that each digit in the top row is greater than the digit in the corresponding bottom row, subtraction can proceed as normal, and the answer is found to be 248.

Multiplication

Multiplication involves adding together multiple copies of a number. It is indicated by an × symbol or a number immediately outside of a parenthesis. For example:

$$5(8-2)$$

The two numbers being multiplied together are called **factors**, and their result is called a **product**. For example, $9 \times 6 = 54$. This can be shown alternatively by expansion of either the 9 or the 6:

$$9 \times 6 = 9 + 9 + 9 + 9 + 9 + 9 = 54$$

$$9 \times 6 = 6 + 6 + 6 + 6 + 6 + 6 + 6 + 6 + 6 = 54$$

Like addition, multiplication holds the commutative and associative properties:

$$115 = 23 \times 5 = 5 \times 23 = 115$$

$$84 = 3 \times (7 \times 4) = (3 \times 7) \times 4 = 84$$

Multiplication also follows the distributive property, which allows the multiplication to be distributed through parentheses. The formula for distribution is $a \times (b + c) = ab + ac$. This is clear after the examples:

$$45 = 5 \times 9 = 5(3 + 6) = (5 \times 3) + (5 \times 6) = 15 + 30 = 45$$

$$20 = 4 \times 5 = 4(10 - 5) = (4 \times 10) - (4 \times 5) = 40 - 20 = 20$$

Multiplication becomes slightly more complicated when multiplying numbers with decimals. The easiest way to answer these problems is to ignore the decimals and multiply as if they were whole numbers. After multiplying the factors, place a decimal in the product. The placement of the decimal is determined by taking the cumulative number of decimal places in the factors.

For example:

0.7	2.6	1.5
x 3	x 4.2	x 6.4
2.1	10.92	9.60

Let's tackle the first example. First, ignore the decimal and multiply the numbers as though they were whole numbers to arrive at a product: 21. Second, count the number of digits that follow a decimal (one). Finally, move the decimal place that many positions to the left, as the factors have only one decimal place. The second example works the same way, except that there are two total decimal places in the factors, so the product's decimal is moved two places over. In the third example, the decimal should be moved over two digits, but the digit zero is no longer needed, so it is erased, and the final answer is 9.6.

Division

Division and multiplication are inverses of each other in the same way that addition and subtraction are opposites. The signs designating a division operation are the ÷ and / symbols. In division, the second number divides into the first. The number before the division sign is called the **dividend** or, if expressed as a fraction, the **numerator**. For example, in $a \div b$, a is the dividend, while in $\frac{a}{b}$, a is the numerator.

The number after the division sign is called the **divisor** or, if expressed as a fraction, the **denominator**. For example, in $a \div b$, b is the divisor, while in $\frac{a}{b}$, b is the denominator.

Like subtraction, division doesn't follow the commutative property, as it matters which number comes before the division sign, and division doesn't follow the associative or distributive properties for the same reason. For example:

$$\frac{3}{2} = 9 \div 6 \neq 6 \div 9 = \frac{2}{3}$$

$$2 = 10 \div 5 = (30 \div 3) \div 5 \neq 30 \div (3 \div 5) = 30 \div \frac{3}{5} = 50$$

$$25 = 20 + 5 = (40 \div 2) + (40 \div 8) \neq 40 \div (2 + 8) = 40 \div 10 = 4$$

If a divisor doesn't divide into a dividend an integer number of times, whatever is left over is termed the **remainder**. The remainder can be further divided out into decimal form by using long division; however, this doesn't always give a quotient with a finite number of decimal places, so the remainder can also be expressed as a fraction over the original divisor.

Division with decimals is similar to multiplication with decimals in that when dividing a decimal by a whole number, ignore the decimal and divide as if it were a whole number.

Upon finding the answer, or quotient, place the decimal at the decimal place equal to that in the dividend.

$$15.75 \div 3 = 5.25$$

When the divisor is a decimal number, multiply both the divisor and dividend by 10. Repeat this until the divisor is a whole number, then complete the division operation as described above.

$$17.5 \div 2.5 = 175 \div 25 = 7$$

Order of Operations

When working with complicated expressions, parentheses are used to indicate in which order to perform operations. However, to avoid having too many parentheses in an expression, here are some basic rules concerning the proper order to perform operations when not otherwise specified.

1. Parentheses: always perform operations inside parentheses first, regardless of what those operations are
2. Exponents
3. Multiplication and Division
4. Addition and Subtraction

For #3 & #4, work these from left to right. So, if there a subtraction problem and then an addition problem, the subtraction problem will be worked first.

Note that multiplication and division are performed from left to right as they appear in the expression or equation. Addition and subtraction also are performed from left to right as they appear.

As an aid to memorizing this, some students like to use the mnemonic PEMDAS. Furthermore, this acronym can be associated with a mnemonic phrase such as "Pirates Eat Many Donuts At Sea."

Evaluate the following two problems to understand the Order of Operations:

1) $4 + (3 \times 2)^2 \div 4$

First, solve the operation within the parentheses: $4 + 6^2 \div 4$.
Second, solve the exponent: $4 + 36 \div 4$.
Third, solve the division operation: $4 + 9$.
Fourth, finish the operation with addition for the answer, 13.

2) $2 \times (6 + 3) \div (2 + 1)^2$

$2 \times 9 \div (3)^2$
$2 \times 9 \div 9$
$18 \div 9$
2

Properties of Operations

Properties of operations exist that make calculations easier and solve problems for missing values. The following table summarizes commonly used properties of real numbers.

Property	**Addition**	**Multiplication**
Commutative	$a + b = b + a$	$a \times b = b \times a$
Associative	$(a + b) + c = a + (b + c)$	$(a \times b) \times c = a \times (b \times c)$
Identity	$a + 0 = a;\ 0 + a = a$	$a \times 1 = a;\ 1 \times a = a$
Inverse	$a + (-a) = 0$	$a \times \frac{1}{a} = 1;\ a \neq 0$
Distributive	$a(b + c) = ab + ac$	

The cumulative property of addition states that the order in which numbers are added does not change the sum. Similarly, the commutative property of multiplication states that the order in which numbers are multiplied does not change the product. The associative property of addition and multiplication state that the grouping of numbers being added or multiplied does not change the sum or product, respectively. The commutative and associative properties are useful for performing calculations. For example, $(47 + 25) + 3$ is equivalent to $(47 + 3) + 25$, which is easier to calculate.

The identity property of addition states that adding zero to any number does not change its value. The identity property of multiplication states that multiplying a number by 1 does not change its value. The inverse property of addition states that the sum of a number and its opposite equals zero. Opposites are numbers that are the same with different signs (ex. 5 and -5; $-\frac{1}{2}$ and $\frac{1}{2}$). The inverse property of multiplication states that the product of a number (other than 0) and its reciprocal equals 1. Reciprocal numbers have numerators and denominators that are inverted (ex. $\frac{2}{5}$ and $\frac{5}{2}$). Inverse properties are useful for canceling quantities to find missing values (see algebra content). For example, $a + 7 = 12$ is solved by adding the inverse of 7 (which is -7) to both sides in order to isolate a.

The distributive property states that multiplying a sum (or difference) by a number produces the same result as multiplying each value in the sum (or difference) by the number and adding (or subtracting) the products. Consider the following scenario: You are buying three tickets for a baseball game. Each ticket costs $18. You are also charged a fee of $2 per ticket for purchasing the tickets online. The cost is calculated: $3 \times 18 + 3 \times 2$. Using the distributive property, the cost can also be calculated $3(18 + 2)$.

Comparing, Classifying, and Ordering Real Numbers

A **rational number** is any number that can be written as a fraction or ratio. Within the set of rational numbers, several subsets exist that are referenced throughout the mathematics topics. **Counting numbers** are the first numbers learned as a child. Counting numbers consist of 1,2,3,4, and so on. **Whole numbers** include all counting numbers and zero (0,1,2,3,4,…). **Integers** include counting numbers, their opposites, and zero (…,-3,-2,-1,0,1,2,3,…). **Rational numbers** are inclusive of integers, fractions, and decimals that terminate, or end (1.7, 0.04213) or repeat ($0.136\overline{5}$).

A **number line** typically consists of integers (…3,2,1,0,-1,-2,-3…), and is used to visually represent the value of a rational number. Each rational number has a distinct position on the line determined by comparing its value with the displayed values on the line. For example, if plotting -1.5 on the number line below, it is necessary to recognize that the value of -1.5 is .5 less than -1 and .5 greater than -2. Therefore, -1.5 is plotted halfway between -1 and -2.

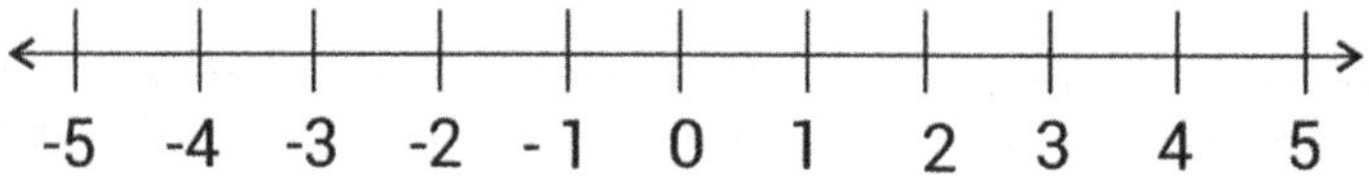

The number system that is used consists of only ten different digits or characters. However, this system is used to represent an infinite number of values. The **place value system** makes this infinite number of values possible. The position in which a digit is written corresponds to a given value. Starting from the decimal point (which is implied, if not physically present), each subsequent place value to the left represents a value greater than the one before it. Conversely, starting from the decimal point, each subsequent place value to the right represents a value less than the one before it.

The names for the place values to the left of the decimal point are as follows:

…	Billions	Hundred-Millions	Ten-Millions	Millions	Hundred-Thousands	Ten-Thousands	Thousands	Hundreds	Tens	Ones

*Note that this table can be extended infinitely further to the left.

The names for the place values to the right of the decimal point are as follows:

Decimal Point (.)	Tenths	Hundredths	Thousandths	Ten-Thousandths	…

*Note that this table can be extended infinitely further to the right.

When given a multi-digit number, the value of each digit depends on its place value. Consider the number 682,174.953. Referring to the chart above, it can be determined that the digit 8 is in the ten-thousands place. It is in the fifth place to the left of the decimal point. Its value is 8 ten-thousands or 80,000. The digit 5 is two places to the right of the decimal point. Therefore, the digit 5 is in the hundredths place. Its value is 5 hundredths or $\frac{5}{100}$ (equivalent to .05).

In accordance with the **base-10 system**, the value of a digit increases by a factor of ten each place it moves to the left. For example, consider the number 7. Moving the digit one place to the left (70), increases its value by a factor of 10 ($7 \times 10 = 70$). Moving the digit two places to the left (700) increases its value by a factor of 10 twice ($7 \times 10 \times 10 = 700$). Moving the digit three places to the left (7,000) increases its value by a factor of 10 three times ($7 \times 10 \times 10 \times 10 = 7{,}000$), and so on.

Conversely, the value of a digit decreases by a factor of ten each place it moves to the right. (Note that multiplying by $\frac{1}{10}$ is equivalent to dividing by 10). For example, consider the number 40. Moving the digit one place to the right (4) decreases its value by a factor of 10 ($40 \div 10 = 4$). Moving the digit two places to the right (0.4), decreases its value by a factor of 10 twice ($40 \div 10 \div 10 = 0.4$) or ($40 \times \frac{1}{10} \times \frac{1}{10} = 0.4$). Moving the digit three places to the right (0.04) decreases its value by a factor of 10 three times ($40 \div 10 \div 10 \div 10 = 0.04$) or ($40 \times \frac{1}{10} \times \frac{1}{10} \times \frac{1}{10} = 0.04$), and so on.

Ordering Numbers

A common question type asks to order rational numbers from least to greatest or greatest to least. The numbers will come in a variety of formats, including decimals, percentages, roots, fractions, and whole numbers. These questions test for knowledge of different types of numbers and the ability to determine their respective values.

Before discussing ordering all numbers, let's start with decimals.

To compare decimals and order them by their value, utilize a method similar to that of ordering large numbers.

The main difference is where the comparison will start. Assuming that any numbers to left of the decimal point are equal, the next numbers to be compared are those immediately to the right of the decimal point. If those are equal, then move on to compare the values in the next decimal place to the right.

For example:

Which number is greater, 12.35 or 12.38?

Check that the values to the left of the decimal point are equal:

12 = 12

Next, compare the values of the decimal place to the right of the decimal:

12.3 = 12.3

Those are also equal in value.

Finally, compare the value of the numbers in the next decimal place to the right on both numbers:

12.3**5** and 12.3**8**

Here the 5 is less than the 8, so the final way to express this inequality is:

$$12.35 < 12.38$$

Comparing decimals is regularly exemplified with money because the "cents" portion of money ends in the hundredths place. When paying for gasoline or meals in restaurants, and even in bank accounts, if enough errors are made when calculating numbers to the hundredths place, they can add up to dollars and larger amounts of money over time.

Now that decimal ordering has been explained, let's expand and consider all real numbers. Whether the question asks to order the numbers from greatest to least or least to greatest, the crux of the question is the same—convert the numbers into a common format. Generally, it's easiest to write the numbers as whole numbers and decimals so they can be placed on a number line. Follow these examples to understand this strategy.

1) Order the following rational numbers from greatest to least:

$$\sqrt{36}, 0.65, 78\%, \frac{3}{4}, 7, 90\%, \frac{5}{2}$$

Of the seven numbers, the whole number (7) and decimal (0.65) are already in an accessible form, so concentrate on the other five.

First, the square root of 36 equals 6. (If the test asks for the root of a non-perfect root, determine which two whole numbers the root lies between.) Next, convert the percentages to decimals. A percentage means "per hundred," so this conversion requires moving the decimal point two places to the left, leaving 0.78 and 0.9. Lastly, evaluate the fractions:

$$\frac{3}{4} = \frac{75}{100} = 0.75\,; \frac{5}{2} = 2\frac{1}{2} = 2.5$$

Now, the only step left is to list the numbers in the request order:

$$7, \sqrt{36}, \frac{5}{2}, 90\%, 78\%, \frac{3}{4}, 0.65$$

2) Order the following rational numbers from least to greatest:

$$2.5, \sqrt{9}, -10.5, 0.853, 175\%, \sqrt{4}, \frac{4}{5}$$

$$\sqrt{9} = 3$$

$$175\% = 1.75$$

$$\sqrt{4} = 2$$

$$\frac{4}{5} = 0.8$$

From least to greatest, the answer is:

$$-10.5, \frac{4}{5}, 0.853, 175\%, \sqrt{4}, 2.5, \sqrt{9}$$

Performing Operations Involving Exponents

Exponents are used in mathematics to express a number or variable multiplied by itself a certain number of times. For example, x^3 means x is multiplied by itself three times. In this expression, x is called the **base**, and 3 is the **exponent.** Exponents can be used in more complex problems when they contain fractions and negative numbers.

Fractional exponents can be explained by looking first at the inverse of exponents, which are **roots.** Given the expression x^2, the square root can be taken, $\sqrt{x^2}$, cancelling out the 2 and leaving *x* by itself, if *x* is positive. Cancellation occurs because $\sqrt{x}$ can be written with exponents, instead of roots, as $x^{\frac{1}{2}}$. The numerator of 1 is the exponent, and the denominator of 2 is called the **root** (which is why it's referred to as a **square root**). Taking the square root of x^2 is the same as raising it to the $\frac{1}{2}$ power. Written out in mathematical form, it takes the following progression:

$$\sqrt{x^2} = (x^2)^{\frac{1}{2}} = x$$

From properties of exponents, $2 \times \frac{1}{2} = 1$ is the actual exponent of *x*. Another example can be seen with $x^{\frac{4}{7}}$. The variable *x,* raised to four-sevenths, is equal to the seventh root of *x* to the fourth power: $\sqrt[7]{x^4}$. In general,

$$x^{\frac{1}{n}} = \sqrt[n]{x}$$

and

$$x^{\frac{m}{n}} = \sqrt[n]{x^m}$$

Negative exponents also involve fractions. Whereas y^3 can also be rewritten as $\frac{y^3}{1}$, y^{-3} can be rewritten as $\frac{1}{y^3}$. A negative exponent means the exponential expression must be moved to the opposite spot in a fraction to make the exponent positive. If the negative appears in the numerator, it moves to the denominator. If the negative appears in the denominator, it is moved to the numerator. In general, $a^{-n} = \frac{1}{a^n}$, and a^{-n} and a^n are reciprocals.

Take, for example, the following expression:

$$\frac{a^{-4}b^2}{c^{-5}}$$

Since *a* is raised to the negative fourth power, it can be moved to the denominator. Since *c* is raised to the negative fifth power, it can be moved to the numerator. The *b* variable is raised to the positive second power, so it does not move.

The simplified expression is as follows:

$$\frac{b^2c^5}{a^4}$$

In mathematical expressions containing exponents and other operations, the order of operations must be followed. **PEMDAS** states that exponents are calculated after any parenthesis and grouping symbols but before any multiplication, division, addition, and subtraction.

Simplifying and Approximating Radicals

The **square root symbol** is expressed as $\sqrt{}$ and is commonly known as the **radical.** Taking the root of a number is the inverse operation of multiplying that number by itself some amount of times. For example, squaring the number 7 is equal to 7×7, or 49. Finding the square root is the opposite of finding an exponent, as the operation seeks a number that when multiplied by itself equals the number in the square root symbol.

For example, $\sqrt{36}$ = 6 because 6 multiplied by 6 equals 36. Note, the square root of 36 is also -6 since -6 x -6 = 36. This can be indicated using a plus/minus symbol like this: ±6. However, square roots are often just expressed as a positive number for simplicity with it being understood that the true value can be either positive or negative.

Perfect squares are squares of whole numbers. The list of perfect squares begins with 0, 1, 4, 9, 16, 25, 36, 49, 64, 81, and 100.

Determining the square root of imperfect squares requires a calculator to reach an exact figure. It's possible, however, to approximate the answer by finding the two perfect squares that the number fits between. For example, the square root of 40 is between 6 and 7 since the squares of those numbers are 36 and 49, respectively.

Square roots are the most common root operation. If the radical doesn't have a number to the upper left of the symbol $\sqrt{}$, then it's a square root. Sometimes a radical includes a number in the upper left, like $\sqrt[3]{27}$, as in the other common root type—the cube root. Complicated roots like the cube root often require a calculator.

Scientific Notation

Scientific Notation is used to represent numbers that are either very small or very large. For example, the distance to the Sun is approximately 150,000,000,000 meters. Instead of writing this number with so many zeros, it can be written in scientific notation as 1.5×10^{11}meters. The same is true for very small numbers, but the exponent becomes negative. If the mass of a human cell is 0.000000000001 kilograms, that measurement can be easily represented by 1.0×10^{-12} kilograms. In both situations, scientific notation makes the measurement easier to read and understand. Each number is translated to an expression with one digit in the tens place multiplied by an expression corresponding to the zeros.

When two measurements are given and both involve scientific notation, it is important to know how these interact with each other:

- In addition and subtraction, the exponent on the ten must be the same before any operations are performed on the numbers. For example, $(1.3 \times 10^4) + (3.0 \times 10^3)$ cannot be added until one of the exponents on the ten is changed. The 3.0×10^3 can be changed to 0.3×10^4, then the 1.3 and 0.3 can be added. The answer comes out to be 1.6×10^4.

- For multiplication, the first numbers can be multiplied and then the exponents on the tens can be added. Once an answer is formed, it may have to be converted into scientific notation again depending on the change that occurred.

- The following is an example of multiplication with scientific notation:

$$(4.5 \times 10^3) \times (3.0 \times 10^{-5}) = 13.5 \times 10^{-2}$$

- Since this answer is not in scientific notation, the decimal is moved over to the left one unit, and 1 is added to the ten's exponent. This results in the final answer: 1.35×10^{-1}.

- For division, the first numbers are divided, and the exponents on the tens are subtracted. Again, the answer may need to be converted into scientific notation form, depending on the type of changes that occurred during the problem.

- **Order of magnitude** relates to scientific notation and is the total count of powers of 10 in a number. For example, there are 6 orders of magnitude in 1,000,000. If a number is raised by an order of magnitude, it is multiplied by 10. Order of magnitude can be helpful in estimating results using very large or small numbers. An answer should make sense in terms of its order of magnitude.

- For example, if area is calculated using two dimensions with 6 orders of magnitude, because area involves multiplication, the answer should have around 12 orders of magnitude. Also, answers can be estimated by rounding to the largest place value in each number. For example, 5,493,302 ×2,523,100 can be estimated by 5 × 3 = 15 with 6 orders of magnitude.

Fractions, Decimals, and Percents

Converting Fractions, Decimals, and Percents

Fractions

A **fraction** is an equation that represents a part of a whole but can also be used to present ratios or division problems. An example of a fraction is $\frac{x}{y}$. In this example, x is called the **numerator,** while y is the **denominator**. The numerator represents the number of parts, and the denominator is the total number of parts. They are separated by a line or slash, known as a fraction bar. In simple fractions, the numerator and denominator can be nearly any integer. However, the denominator of a fraction can never be zero because dividing by zero is a function that is undefined.

Imagine that an apple pie has been baked for a holiday party, and the full pie has eight slices. After the party, there are five slices left. How could the amount of the pie that remains be expressed as a fraction? The numerator is 5 since there are 5 pieces left, and the denominator is 8 since there were eight total slices in the whole pie. Thus, expressed as a fraction, the leftover pie totals $\frac{5}{8}$ of the original amount.

Fractions come in three different varieties: proper fractions, improper fractions, and mixed numbers. **Proper fractions** have a numerator less than the denominator, such as $\frac{3}{8}$, but **improper fractions** have a numerator greater than the denominator, such as $\frac{15}{8}$. **Mixed numbers** combine a whole number with a proper fraction, such as $3\frac{1}{2}$. Any mixed number can be written as an improper fraction by multiplying the integer by the denominator, adding the product to the value of the numerator, and dividing the sum by the original denominator. For example, $3\frac{1}{2} = \frac{3\times2+1}{2} = \frac{7}{2}$. Whole numbers can also be converted into fractions by placing the whole number as the numerator and making the denominator 1. For example, $3 = \frac{3}{1}$.

One of the most fundamental concepts of fractions is their ability to be manipulated by multiplication or division. This is possible since $\frac{n}{n} = 1$ for any non-zero integer. As a result, multiplying or dividing by $\frac{n}{n}$ will not alter the original fraction since any number multiplied or divided by 1 doesn't change the value of that number. Fractions of the same value are known as equivalent fractions. For example, $\frac{2}{4}, \frac{4}{8}, \frac{50}{100}$, and $\frac{75}{150}$ are equivalent, as they all equal $\frac{1}{2}$.

Although many equivalent fractions exist, they are easier to compare and interpret when reduced or simplified. The numerator and denominator of a simple fraction will have no factors in common other than 1. When reducing or simplifying fractions, divide the numerator and denominator by the greatest common factor. A simple strategy is to divide the numerator and denominator by low numbers, like 2, 3, or 5 until arriving at a simple fraction, but the same thing could be achieved by determining the greatest common factor for both the numerator and denominator and dividing each by it. Using the first method is preferable when both the numerator and denominator are even, end in 5, or are obviously a multiple of another number. However, if no numbers seem to work, it will be necessary to factor the numerator and denominator to find the GCF. Let's look at examples:

1) Simplify the fraction $\frac{6}{8}$:

Dividing the numerator and denominator by 2 results in $\frac{3}{4}$, which is a simple fraction.

2) Simplify the fraction $\frac{12}{36}$:

Dividing the numerator and denominator by 2 leaves $\frac{6}{18}$. This isn't a simple fraction, as both the numerator and denominator have factors in common. Dividing each by 3 results in $\frac{2}{6}$, but this can be further simplified by dividing by 2 to get $\frac{1}{3}$. This is the simplest fraction, as the numerator is 1. In cases like this, multiple division operations can be avoided by determining the greatest common factor between the numerator and denominator.

3) Simplify the fraction $\frac{18}{54}$ by dividing by the greatest common factor:

First, determine the factors for the numerator and denominator. The factors of 18 are 1, 2, 3, 6, 9, and 18. The factors of 54 are 1, 2, 3, 6, 9, 18, 27, and 54. Thus, the greatest common factor is 18. Dividing $\frac{18}{54}$

by 18 leaves $\frac{1}{3}$, which is the simplest fraction. This method takes slightly more work, but it definitively arrives at the simplest fraction.

Of the four basic operations that can be performed on fractions, the one that involves the least amount of work is multiplication. To multiply two fractions, simply multiply the numerators together, multiply the denominators together, and place the products of each as a fraction. Whole numbers and mixed numbers can also be expressed as a fraction, as described above, to multiply with a fraction. Let's work through a couple of examples.

$$1)\frac{2}{5} \times \frac{3}{4} = \frac{6}{20} = \frac{3}{10}$$

$$2)\frac{4}{9} \times \frac{7}{11} = \frac{28}{99}$$

Dividing fractions is similar to multiplication with one key difference. To divide fractions, flip the numerator and denominator of the second fraction, and then proceed as if it were a multiplication problem:

$$1)\frac{7}{8} \div \frac{4}{5} = \frac{7}{8} \times \frac{5}{4} = \frac{35}{32}$$

$$2)\frac{5}{9} \div \frac{1}{3} = \frac{5}{9} \times \frac{3}{1} = \frac{15}{9} = \frac{5}{3}$$

Addition and subtraction require more steps than multiplication and division, as these operations require the fractions to have the same denominator, also called a common denominator. It is always possible to find a common denominator by multiplying the denominators. However, when the denominators are large numbers, this method is unwieldy, especially if the answer must be provided in its simplest form. Thus, it's beneficial to find the least common denominator of the fractions—the least common denominator is incidentally also the least common multiple.

Once equivalent fractions have been found with common denominators, simply add or subtract the numerators to arrive at the answer:

$$1)\frac{1}{2} + \frac{3}{4} = \frac{2}{4} + \frac{3}{4} = \frac{5}{4}$$

$$2)\frac{3}{12} + \frac{11}{20} = \frac{15}{60} + \frac{33}{60} = \frac{48}{60} = \frac{4}{5}$$

$$3)\frac{7}{9} - \frac{4}{15} = \frac{35}{45} - \frac{12}{45} = \frac{23}{45}$$

$$4)\frac{5}{6} - \frac{7}{18} = \frac{15}{18} - \frac{7}{18} = \frac{8}{18} = \frac{4}{9}$$

<u>Decimals</u>

The **decimal system** is a way of writing out numbers that uses ten different numerals: 0, 1, 2, 3, 4, 5, 6, 7, 8, and 9. This is also called a "base ten" or "base 10" system. Other bases are also used. For example, computers work with a base of 2. This means they only use the numerals 0 and 1.

The **decimal place** denotes how far to the right of the decimal point a numeral is. The first digit to the right of the decimal point is in the *tenths* place. The next is the **hundredths**. The third is the **thousandths**.

So, 3.142 has a 1 in the tenths place, a 4 in the hundredths place, and a 2 in the thousandths place.

The **decimal point** is a period used to separate the **ones** place from the **tenths** place when writing out a number as a decimal.

A **decimal number** is a number written out with a decimal point instead of as a fraction, for example, 1.25 instead of $\frac{5}{4}$. Depending on the situation, it can sometimes be easier to work with fractions and sometimes easier to work with decimal numbers.

A decimal number is **terminating** if it stops at some point. It is called **repeating** if it never stops but repeats a pattern over and over. It is important to note that every rational number can be written as a terminating decimal or as a repeating decimal.

Addition with Decimals

To add decimal numbers, each number needs to be lined up by the decimal point in vertical columns. For each number being added, the zeros to the right of the last number need to be filled in so that each of the numbers has the same number of places to the right of the decimal. Then, the columns can be added together. Here is an example of 2.45 + 1.3 + 8.891 written in column form:

$$\begin{array}{r} 2.450 \\ 1.300 \\ \underline{+\ 8.891} \end{array}$$

Zeros have been added in the columns so that each number has the same number of places to the right of the decimal.

Added together, the correct answer is 12.641:

$$\begin{array}{r} 2.450 \\ 1.300 \\ \underline{+\ 8.891} \\ 12.641 \end{array}$$

Subtraction with Decimals

Subtracting decimal numbers is the same process as adding decimals. Here is 7.89 – 4.235 written in column form:

$$\begin{array}{r} 7.890 \\ -\ \underline{4.235} \\ 3.655 \end{array}$$

A zero has been added in the column so that each number has the same number of places to the right of the decimal.

Multiplication with Decimals
The simplest way to multiply decimals is to calculate the product as if the decimals are not there, then count the number of decimal places in the original problem. Use that total to place the decimal the same number of places over in your answer, counting from right to left. For example, 0.5 x 1.25 can be rewritten and multiplied as 5 x 125, which equals 625. Then the decimal is added three places from the right for .625.

The final answer will have the same number of decimal *points* as the total number of decimal *places* in the problem. The first number has one decimal place, and the second number has two decimal places. Therefore, the final answer will contain three decimal places:

$$0.5 \times 1.25 = 0.625$$

Division with Decimals
Dividing a decimal by a whole number entails using long division first by ignoring the decimal point. Then, the decimal point is moved the number of places given in the problem.

For example, $6.8 \div 4$ can be rewritten as $68 \div 4$, which is 17. There is one non-zero integer to the right of the decimal point, so the final solution would have one decimal place to the right of the solution. In this case, the solution is 1.7.

Dividing a decimal by another decimal requires changing the divisor to a whole number by moving its decimal point. The decimal place of the dividend should be moved by the same number of places as the divisor. Then, the problem is the same as dividing a decimal by a whole number.

For example, $5.72 \div 1.1$ has a divisor with one decimal point in the denominator. The expression can be rewritten as $57.2 \div 11$ by moving each number one decimal place to the right to eliminate the decimal. The long division can be completed as $572 \div 11$ with a result of 52. Since there is one non-zero integer to the right of the decimal point in the problem, the final solution is 5.2.

In another example, $8 \div 0.16$ has a divisor with two decimal points in the denominator. The expression can be rewritten as $800 \div 16$ by moving each number two decimal places to the right to eliminate the decimal in the divisor. The long division can be completed with a result of 50.

<u>Percentages</u>
Think of percentages as fractions with a denominator of 100. In fact, percentage means "per hundred." Problems often require converting numbers from percentages, fractions, and decimals.

The basic percent equation is the following:

$$\frac{is}{of} = \frac{\%}{100}$$

The placement of numbers in the equation depends on what the question asks.

Example 1
Find 40% of 80.

Basically, the problem is asking, "What is 40% of 80?" The 40% is the percent, and 80 is the number to find the percent "of."

The equation is:

$$\frac{x}{80} = \frac{40}{100}$$

After cross-multiplying, the problem becomes 100x = 80(40). Solving for x gives the answer: x = 32.

Example 2
What percent of 100 is 20?

The 20 fills in the "is" portion, while 100 fills in the "of." The question asks for the percent, so that will be x, the unknown. The following equation is set up:

$$\frac{20}{100} = \frac{x}{100}$$

Cross-multiplying yields the equation 100x = 20(100). Solving for x gives the answer of 20%.

Example 3
30% of what number is 30?

The following equation uses the clues and numbers in the problem:

$$\frac{30}{x} = \frac{30}{100}$$

Cross-multiplying results in the equation 30(100) = 30x. Solving for x gives the answer x = 100.

<u>Conversions</u>
Decimals and Percentages
Since a percentage is based on "per hundred," decimals and percentages can be converted by multiplying or dividing by 100. Practically speaking, this always amounts to moving the decimal point two places to the right or left, depending on the conversion. To convert a percentage to a decimal, move the decimal point two places to the left and remove the % sign. To convert a decimal to a percentage, move the decimal point two places to the right and add a "%" sign. Here are some examples:

65% = 0.65
0.33 = 33%
0.215 = 21.5%
99.99% = 0.9999
500% = 5.00
7.55 = 755%

Fractions and Percentages
Remember that a percentage is a number per one hundred. So a percentage can be converted to a fraction by making the number in the percentage the numerator and putting 100 as the denominator:

$$43\% = \frac{43}{100}$$

$$97\% = \frac{97}{100}$$

Note that the percent symbol (%) kind of looks like a 0, a 1, and another 0. So think of a percentage like 54% as 54 over 100.

To convert a fraction to a percent, follow the same logic. If the fraction happens to have 100 in the denominator, you're in luck. Just take the numerator and add a percent symbol:

$$\frac{28}{100} = 28\%$$

Otherwise, divide the numerator by the denominator to get a decimal:

$$\frac{9}{12} = 0.75$$

Then convert the decimal to a percentage:

$$0.75 = 75\%$$

Another option is to make the denominator equal to 100. Be sure to multiply the numerator and the denominator by the same number. For example:

$$\frac{3}{20} \times \frac{5}{5} = \frac{15}{100}$$

$$\frac{15}{100} = 15\%$$

Changing Fractions to Decimals

To change a fraction into a decimal, divide the denominator into the numerator until there are no remainders. There may be repeating decimals, so rounding is often acceptable. A straight line above the repeating portion denotes that the decimal repeats.

Example

Express 4/5 as a decimal.

Set up the division problem.

$$5\overline{)4}$$

5 does not go into 4, so place the decimal and add a zero.

$$5\overline{)4.0}$$

5 goes into 40 eight times. There is no remainder.

$$\begin{array}{r} 0.8 \\ 5\overline{)4.0} \\ -\,4.0 \\ \hline 0 \end{array}$$

The solution is 0.8.

Example

Express 33 1/3 as a decimal.

Since the whole portion of the number is known, set it aside to calculate the decimal from the fraction portion.

Set up the division problem.

$$3\overline{)1}$$

3 does not go into 1, so place the decimal and add zeros. 3 goes into 10 three times.

$$\begin{array}{r} 0.333 \\ 3\overline{)1.000} \end{array}$$

This will repeat with a remainder of 1, so place a line over the 3 denotes the repetition.

$$\begin{array}{r} 0.333 \\ 3\overline{)1.000} \\ -9 \\ \hline 10 \\ -9 \\ \hline 10 \end{array}$$

The solution is $0.\overline{3}$

Changing Decimals to Fractions

To change decimals to fractions, place the decimal portion of the number, the numerator, over the respective place value, the denominator, then reduce, if possible.

Example: Express 0.25 as a fraction.

This is read as twenty-five hundredths, so put 25 over 100. Then reduce to find the solution.

$$\frac{25}{100} = \frac{1}{4}$$

Example: Express 0.455 as a fraction

This is read as four hundred fifty-five thousandths, so put 455 over 1000. Then reduce to find the solution.

$$\frac{455}{1000} = \frac{91}{200}$$

There are two types of problems that commonly involve percentages. The first is to calculate some percentage of a given quantity, where you convert the percentage to a decimal, and multiply the quantity by that decimal. Secondly, you are given a quantity and told it is a fixed percent of an unknown quantity. In this case, convert to a decimal, then divide the given quantity by that decimal.

Example: What is 30% of 760?

Convert the percent into a useable number. "Of" means to multiply.

$$30\% = 0.30$$

Set up the problem based on the givens, and solve.

$$0.30 \times 760 = 228$$

Example: 8.4 is 20% of what number?

Convert the percent into a useable number.

$$20\% = 0.20$$

The given number is a percent of the answer needed, so divide the given number by this decimal rather than multiplying it.

$$\frac{8.4}{0.20} = 42$$

Representing Fractions, Decimals, and Percents Using Various Models

A **fraction** is a part of something that is whole. Items such as apples can be cut into parts to help visualize fractions. If an apple is cut into 2 equal parts, each part represents ½ of the apple. If each half is cut into two parts, the apple now is cut into quarters. Each piece now represents ¼ of the apple. In this example, each part is equal because they all have the same size. Geometric shapes, such as circles and squares, can also be utilized in the classroom to help visualize the idea of fractions.

For example, a circle can be drawn on the board and divided into 6 equal parts:

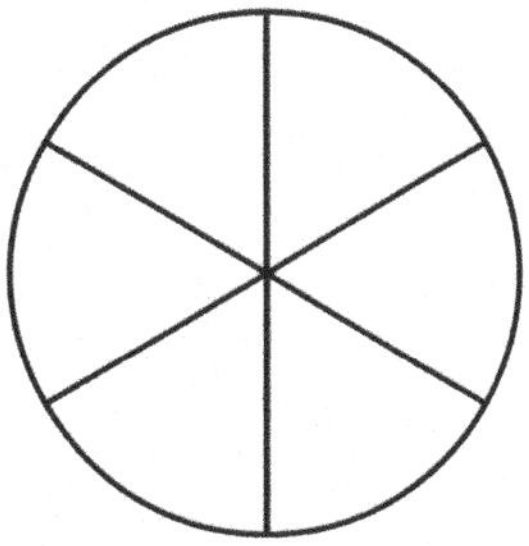

Shading can be used to represent parts of the circle that can be translated into fractions. The top of the fraction, the **numerator,** can represent how many segments are shaded. The bottom of the fraction, the **denominator,** can represent the number of segments that the circle is broken into. A pie is a good analogy to use in this example. If one piece of the circle is shaded, or one piece of pie is cut out, $^1/_6$ of the object is being referred to. An apple, a pie, or a circle can be utilized in order to compare simple fractions. For example, showing that $^1/_2$ is larger than $^1/_4$ and that $^1/_4$ is smaller than $^1/_3$ can be accomplished through shading. A **unit fraction** is a fraction in which the numerator is 1, and the denominator is a positive whole number. It represents one part of a whole—one piece of pie.

When representing fractions, it is important to remember the meaning of the word *fraction*. The number is a fraction, or portion, of a whole. The whole, or 1, is the number on the bottom part of the fraction, the denominator. For example, the fraction $\frac{2}{5}$ represents a number of portions (the numerator, 2) of how many it would take to make a whole (the denominator, 5). So, $\frac{2}{5}$ represents 2 portions out of the 5 it would take to make a whole.

This could also be represented with blocks, as follows:

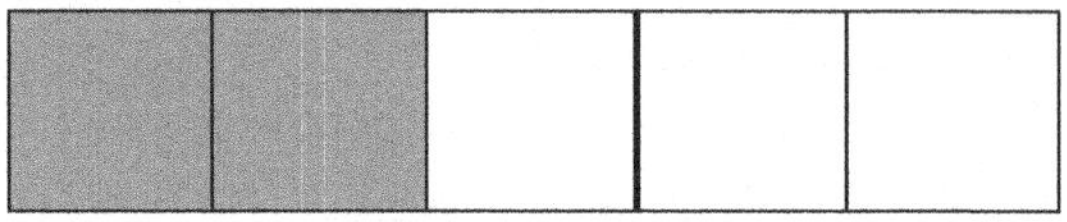

5 blocks = 1 whole, so only 2 of the 5 are shaded.

This method could also be used to represent fractions with a higher number in the numerator than in the denominator.

What does the fraction $\frac{6}{5}$ look like with the block method?

Fractions can be broken apart into sums of fractions with the same denominator. For example, the fraction $\frac{5}{6}$ can be decomposed into sums of fractions with all denominators equal to 6 and the numerators adding to 5. The fraction $\frac{5}{6}$ is decomposed as an of the following:

$$\frac{3}{6}+\frac{2}{6}$$

$$\frac{2}{6}+\frac{2}{6}+\frac{1}{6}$$

$$\frac{3}{6}+\frac{1}{6}+\frac{1}{6}$$

$$\frac{1}{6}+\frac{1}{6}+\frac{1}{6}+\frac{2}{6}$$

$$\frac{1}{6}+\frac{1}{6}+\frac{1}{6}+\frac{1}{6}+\frac{1}{6}$$

As mentioned unit fraction is a fraction in which the numerator is 1. If decomposing a fraction into unit fractions, the sum will consist of a unit fraction added the number of times equal to the numerator. For example, $\frac{3}{4} = \frac{1}{4} + \frac{1}{4} + \frac{1}{4}$ (unit fractions $\frac{1}{4}$ added 3 times). Composing fractions is simply the opposite of decomposing. It is the process of adding fractions with the same denominators to produce a single fraction. For example, $\frac{3}{7} + \frac{2}{7} = \frac{5}{7}$ and $\frac{1}{5} + \frac{1}{5} + \frac{1}{5} = \frac{3}{5}$.

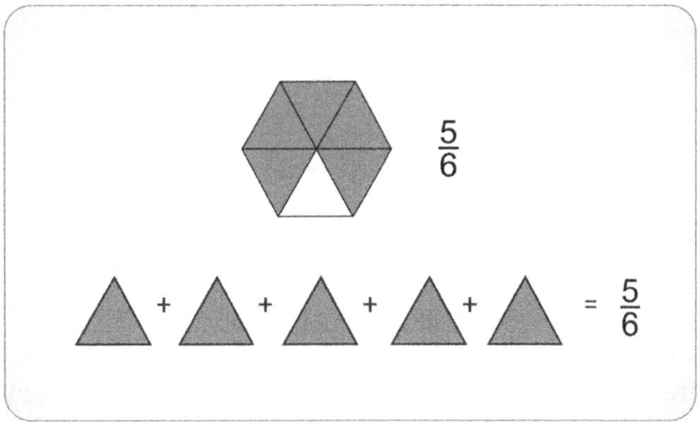

To represent fractions and decimals as distances beginning at zero on a number line, it's helpful to relate the fraction to a real-world application. For example, a charity walk covers $\frac{3}{10}$ of a mile. How could this distance be represented on a number line?

First, divide the number line into tenths, as follows:

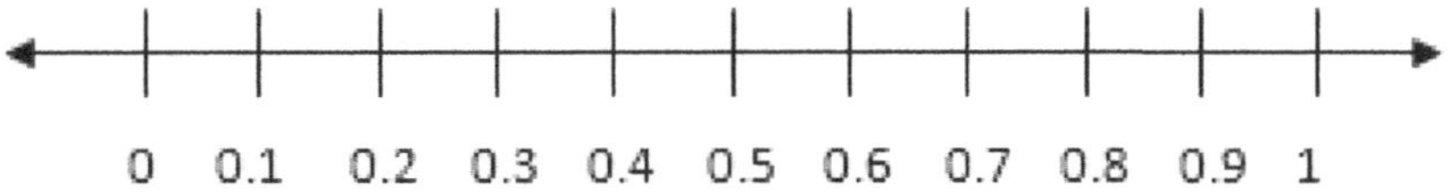

If each division on the number line represents one-tenth of one, or $\frac{1}{10}$, then representing the distance of the charity walk, $\frac{3}{10}$, would cover 3 of those divisions and look as follows:

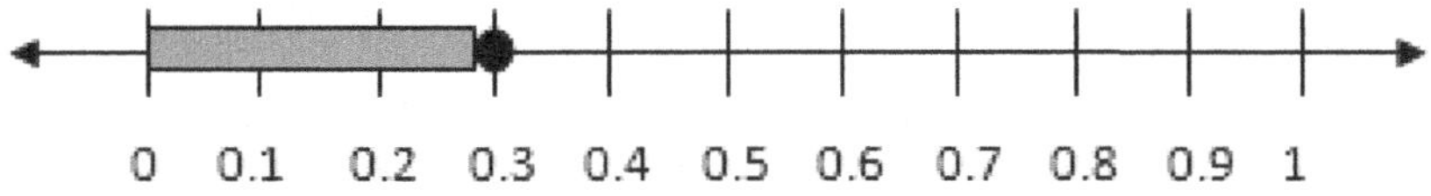

So, the fraction $\frac{3}{10}$ is represented by covering from 0 to 0.3 (or 3 sections) on the number line.

In general, to precisely understand a number being represented on a number line, the first step is to identify how the number line is divided up. When utilizing a number line to represent decimal portions of numbers, it is helpful to label the divisions, or insert additional divisions, as needed.

For example, what number, to the nearest hundredths place is marked by the point on the following number line?

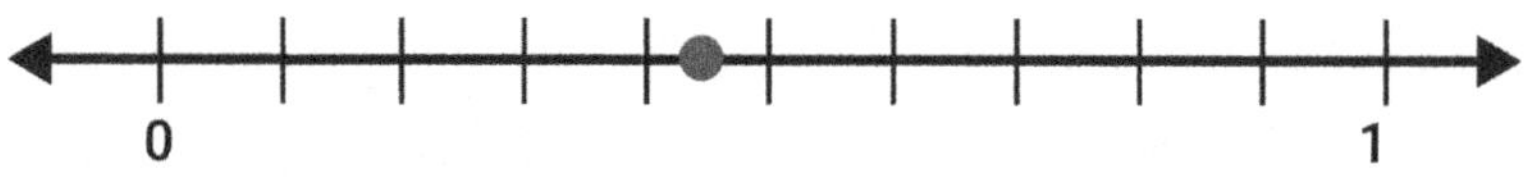

First, figure out how the number line is divided up. In this case, it has ten sections, so it is divided into tenths. To use this number line with the divisions, label the divisions as follows:

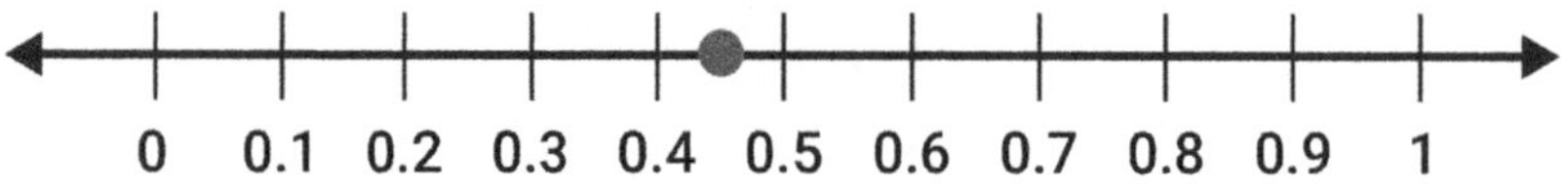

Because the dot is placed equally between 0.4 and 0.5, it is at 0.45.

What number, to the nearest tenths place, is marked by the point on the following number line?

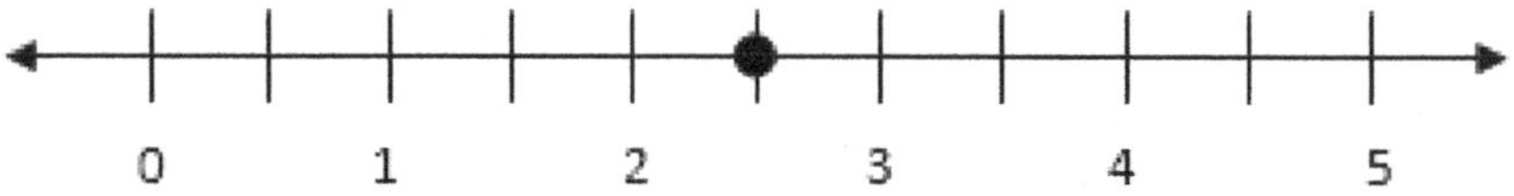

First, determine what the number line is divided up into and mark it on the line.

This number line is divided up into half of the whole numbers it represents, or 0.5 increments. The division and labeling of the number line assists in easily reading the dot as marking 2.5.

Ratio Reasoning

Using Ratios to Describe a Relationship Between Two Quantities

Ratios are used to show the relationship between two quantities. The ratio of oranges to apples in the grocery store may be 3 to 2. That means that for every 3 oranges, there are 2 apples. This comparison can be expanded to represent the actual number of oranges and apples, such as 36 oranges to 24 apples. Another example may be the number of boys to girls in a math class. If the ratio of boys to girls is given as 2 to 5, that means there are 2 boys to every 5 girls in the class. Ratios can also be compared if the units in each ratio are the same. The ratio of boys to girls in the math class can be compared to the ratio of boys to girls in a science class by stating which ratio is higher and which is lower.

Rates are used to compare two quantities with different units. **Unit rates** are the simplest form of rate. With unit rates, the denominator in the comparison of two units is one. For example, if someone can type at a rate of 1000 words in 5 minutes, then his or her unit rate for typing is $\frac{1000}{5} = 200$ words in one minute or 200 words per minute. Any rate can be converted into a unit rate by dividing to make the denominator one. 1000 words in 5 minutes has been converted into the unit rate of 200 words per minute.

Computing Unit Rates

Unit rate word problems will ask to calculate the rate or quantity of something in a different value. For example, a problem might say that a car drove a certain number of miles in a certain number of minutes and then ask how many miles per hour the car was traveling. These questions involve solving proportions. Consider the following examples:

1) Alexandra made $96 during the first 3 hours of her shift as a temporary worker at a law office. She will continue to earn money at this rate until she finishes in 5 more hours. How much does Alexandra make per hour? How much will Alexandra have made at the end of the day?

This problem can be solved in two ways. The first is to set up a proportion, as the rate of pay is constant. The second is to determine her hourly rate, multiply the 5 hours by that rate, and then add the $96.

To set up a proportion, put the money already earned over the hours already worked on one side of an equation. The other side has x over 8 hours (the total hours worked in the day). It looks like this: $\frac{96}{3} = \frac{x}{8}$. Now, cross-multiply to get $768 = 3x$. To get x, divide by 3, which leaves $x = 256$. Alternatively, as x is the numerator of one of the proportions, multiplying by its denominator will reduce the solution by one step. Thus, Alexandra will make $256 at the end of the day. To calculate her hourly rate, divide the total by 8, giving $32 per hour.

Alternatively, it is possible to figure out the hourly rate by dividing $96 by 3 hours to get $32 per hour. Now her total pay can be figured by multiplying $32 per hour by 8 hours, which comes out to $256.

2) Jonathan is reading a novel. So far, he has read 215 of the 335 total pages. It takes Jonathan 25 minutes to read 10 pages, and the rate is constant. How long does it take Jonathan to read one page? How much longer will it take him to finish the novel? Express the answer in time.

To calculate how long it takes Jonathan to read one page, divide the 25 minutes by 10 pages to determine the page per minute rate. Thus, it takes 2.5 minutes to read one page.

Jonathan must read 120 more pages to complete the novel. (This is calculated by subtracting the pages already read from the total.) Now, multiply his rate per page by the number of pages. Thus, $120 \times 2.5 = 300$. Expressed in time, 300 minutes is equal to 5 hours.

3) At a hotel, $\frac{4}{5}$ of the 120 rooms are booked for Saturday. On Sunday, $\frac{3}{4}$ of the rooms are booked. On which day are more of the rooms booked, and by how many more?

The first step is to calculate the number of rooms booked for each day. Do this by multiplying the fraction of the rooms booked by the total number of rooms.

Saturday: $\frac{4}{5} \times 120 = \frac{4}{5} \times \frac{120}{1} = \frac{480}{5} = 96$ rooms

Sunday: $\frac{3}{4} \times 120 = \frac{3}{4} \times \frac{120}{1} = \frac{360}{4} = 90$ rooms

Thus, more rooms were booked on Saturday by 6 rooms.

4) In a veterinary hospital, the veterinarian-to-pet ratio is 1:9. The ratio is always constant. If there are 45 pets in the hospital, how many veterinarians are currently in the veterinary hospital?

Set up a proportion to solve for the number of veterinarians: $\frac{1}{9} = \frac{x}{45}$

Cross-multiplying results in $9x = 45$, which works out to 5 veterinarians.

Alternatively, as there are always 9 times as many pets as veterinarians, it is possible to divide the number of pets (45) by 9. This also arrives at the correct answer of 5 veterinarians.

5) At a general practice law firm, 30% of the lawyers work solely on tort cases. If 9 lawyers work solely on tort cases, how many lawyers work at the firm?

First, solve for the total number of lawyers working at the firm, which will be represented here with x. The problem states that 9 lawyers work solely on torts cases, and they make up 30% of the total lawyers at the firm. Thus, 30% multiplied by the total, x, will equal 9. Written as equation, this is: $30\% \times x = 9$.

It's easier to deal with the equation after converting the percentage to a decimal, leaving $0.3x = 9$. Thus, $x = \frac{9}{0.3} = 30$ lawyers working at the firm.

6) Xavier was hospitalized with pneumonia. He was originally given 35mg of antibiotics. Later, after his condition continued to worsen, Xavier's dosage was increased to 60mg. What was the percent increase of the antibiotics? Round the percentage to the nearest tenth.

An increase or decrease in percentage can be calculated by dividing the difference in amounts by the original amount and multiplying by 100. Written as an equation, the formula is:

$$\frac{new\ quantity\ - old\ quantity}{old\ quantity} \times 100$$

Here, the question states that the dosage was increased from 35mg to 60mg, so these are plugged into the formula to find the percentage increase.

$$\frac{60\ -\ 35}{35} \times 100 = \frac{25}{35} \times 100 = .7142 \times 100 = 71.4\%$$

Using Ratio Reasoning to Convert Rates

Ratios and rates can be used together to convert rates into different units. For example, if someone is driving 50 kilometers per hour, that rate can be converted into miles per hour by using a ratio known as the **conversion factor**. Since the given value contains kilometers and the final answer needs to be in miles, the ratio relating miles to kilometers needs to be used. There are 0.62 miles in 1 kilometer. This, written as a ratio and in fraction form, is $\frac{0.62\ miles}{1\ km}$. To convert 50km/hour into miles per hour, the following conversion needs to be set up: $\frac{50\ km}{hour} \times \frac{0.62\ miles}{1\ km} = 31\ miles\ per\ hour.$

Solving Problems Involving Scale Factors

The ratio between two similar geometric figures is called the **scale factor**. For example, a problem may depict two similar triangles, A and B. The scale factor from the smaller triangle A to the larger triangle B is given as 2 because the length of the corresponding side of the larger triangle, 16, is twice the corresponding side on the smaller triangle, 8. This scale factor can also be used to find the value of a missing side, x, in triangle A. Since the scale factor from the smaller triangle (A) to larger one (B) is 2, the larger corresponding side in triangle B (given as 25), can be divided by 2 to find the missing side in A (x = 12.5). The scale factor can also be represented in the equation $2A = B$ because two times the lengths of A gives the corresponding lengths of B. This is the idea behind similar triangles.

Using Proportional Relationships to Solve Real-World Problems

Proportional Relationships

Much like a scale factor can be written using an equation like $2A = B$, a **relationship** is represented by the equation $Y = kX$. X and Y are **proportional** because as values of X increase, the values of Y also increase. A relationship that is **inversely proportional** can be represented by the equation $Y = \frac{k}{X}$, where the value of Y decreases as the value of x increases and vice versa.

Proportional reasoning can be used to solve problems involving ratios, percentages, and averages. Ratios can be used in setting up proportions and solving them to find unknowns. For example, if a student completes an average of 10 pages of math homework in 3 nights, how long would it take the student to complete 22 pages? Both ratios can be written as fractions. The second ratio would contain the unknown.

The following proportion represents this problem, where x is the unknown number of nights:

$$\frac{10\ pages}{3\ nights} = \frac{22\ pages}{x\ nights}$$

Solving this proportion entails cross-multiplying and results in the following equation: $10x = 22 \times 3$. Simplifying and solving for x results in the exact solution: $x = 6.6\ nights$. The result would be rounded up to 7 because the homework would actually be completed on the 7th night.

The following problem uses ratios involving percentages:

If 20% of the class is girls and 30 students are in the class, how many girls are in the class?

To set up this problem, it is helpful to use the common proportion: $\frac{\%}{100} = \frac{is}{of}$. Within the proportion, % is the percentage of girls, 100 is the total percentage of the class, *is* is the number of girls, and *of* is the total number of students in the class. Most percentage problems can be written using this language. To solve this problem, the proportion should be set up as $\frac{20}{100} = \frac{x}{30}$, and then solved for x. Cross-multiplying results in the equation $20 \times 30 = 100x$, which results in the solution $x = 6$. There are 6 girls in the class.

Problems involving volume, length, and other units can also be solved using ratios. For example, a problem may ask for the volume of a cone to be found that has a radius, $r = 7m$ and a height, $h = 16m$. Referring to the formulas provided on the test, the volume of a cone is given as: $V = \pi r^2 \frac{h}{3}$, where r is the radius, and h is the height. Plugging $r = 7$ and $h = 16$ into the formula, the following is obtained: $V = \pi(7^2)\frac{16}{3}$. Therefore, volume of the cone is found to be approximately $821m^3$. Sometimes, answers in different units are sought. If this problem wanted the answer in liters, $821m^3$ would need to be converted. Using the equivalence statement $1m^3 = 1000L$, the following ratio would be used to solve for liters: $821\text{m}^3 \times \frac{1000L}{1m^3}$. Cubic meters in the numerator and denominator cancel each other out, and the answer is converted to 821,000 liters, or 8.21×10^5 L.

Other conversions can also be made between different given and final units. If the temperature in a pool is 30°C, what is the temperature of the pool in degrees Fahrenheit? To convert these units, an equation is used relating Celsius to Fahrenheit. The following equation is used: $T_{°F} = 1.8T_{°C} + 32$. Plugging in the given temperature and solving the equation for T yields the result: $T_{°F} = 1.8(30) + 32 = 86°F$. Both units in the metric system and U.S. customary system are widely used.

Here are some more examples of how to solve for proportions:

1) $\frac{75\%}{90\%} = \frac{25\%}{x}$

To solve for x, the fractions must be cross multiplied: ($75\%x = 90\% \times 25\%$). To make things easier, let's convert the percentages to decimals: ($0.9 \times 0.25 = 0.225 = 0.75x$). To get rid of x's coefficient, each side must be divided by that same coefficient to get the answer $x = 0.3$. The question could ask for the answer as a percentage or fraction in lowest terms, which are 30% and $\frac{3}{10}$, respectively.

2) $\frac{x}{12} = \frac{30}{96}$

Cross-multiply: $96x = 30 \times 12$

Multiply: $96x = 360$

Divide: $x = 360 \div 96$

Answer: $x = 3.75$

3) $\frac{0.5}{3} = \frac{x}{6}$

Cross-multiply: $3x = 0.5 \times 6$

Multiply: $3x = 3$

Divide: $x = 3 \div 3$

Answer: $x = 1$

You may have noticed there's a faster way to arrive at the answer. If there is an obvious operation being performed on the proportion, the same operation can be used on the other side of the proportion to solve for x. For example, in the first practice problem, 75% became 25% when divided by 3, and upon doing the same to 90%, the correct answer of 30% would have been found with much less legwork. However, these questions aren't always so intuitive, so it's a good idea to work through the steps, even if the answer seems apparent from the outset.

Solving Ratio and Percent Problems

Questions dealing with percentages can be difficult when they are phrased as word problems. These word problems almost always come in three varieties. The first type will ask to find what percentage of some number will equal another number. The second asks to determine what number is some percentage of another given number. The third will ask what number another number is a given percentage of.

One of the most important parts of correctly answering percentage word problems is to identify the numerator and the denominator. This fraction can then be converted into a percentage, as described above.

The following word problem shows how to make this conversion:

A department store carries several different types of footwear. The store is currently selling 8 athletic shoes, 7 dress shoes, and 5 sandals. What percentage of the store's footwear are sandals?

First, calculate what serves as the 'whole,' as this will be the denominator. How many total pieces of footwear does the store sell? The store sells 20 different types (8 athletic + 7 dress + 5 sandals).

Second, what footwear type is the question specifically asking about? Sandals. Thus, 5 is the numerator.

Third, the resultant fraction must be expressed as a percentage. The first two steps indicate that $\frac{5}{20}$ of the footwear pieces are sandals. This fraction must now be converted into a percentage:

$$\frac{5}{20} \times \frac{5}{5} = \frac{25}{100} = 25\%$$

Basic Concepts of Number Theory

Prime and Composite Numbers

Whole numbers are classified as either prime or composite. A prime number can only be divided evenly by itself and one. For example, the number 11 can only be divided evenly by 11 and one; therefore, 11 is a prime number. A helpful way to visualize a prime number is to use concrete objects and try to divide them into equal piles. If dividing 11 coins, the only way to divide them into equal piles is to create 1 pile of 11 coins or to create 11 piles of 1 coin each. Other examples of prime numbers include 2, 3, 5, 7, 13, 17, and 19.

A composite number is any whole number that is not a prime number. A composite number is a number that can be divided evenly by one or more numbers other than itself and one. For example, the number 6 can be divided evenly by 2 and 3. Therefore, 6 is a composite number. If dividing 6 coins into equal piles, the possibilities are 1 pile of 6 coins, 2 piles of 3 coins, 3 piles of 2 coins, or 6 piles of 1 coin. Other examples of composite numbers include 4, 8, 9, 10, 12, 14, 15, 16, 18, and 20.

To determine if a number is a prime or composite number, the number is divided by every whole number greater than one and less than its own value. If it divides evenly by any of these numbers, then the number is composite. If it does not divide evenly by any of these numbers, then the number is prime. For example, when attempting to divide the number 5 by 2, 3, and 4, none of these numbers divide evenly. Therefore, 5 must be a prime number.

Odd and Even Numbers

Even numbers are all divisible by the number 2. **Odd numbers** are not divisible by 2, and an odd quantity of items cannot be paired up into groups of 2 without having 1 item leftover. Examples of even numbers are 2, 4, 6, 20, 100, 242, etc. Examples of odd numbers are 1, 3, 5, 27, 99, 333, etc.

Factors, Multiples, and Divisibility

The **Fundamental Theorem of Arithmetic** states that any integer greater than 1 is either a prime number or can be written as a unique product of prime numbers. Factors can be used to find the combination of numbers to multiply to produce an integer that is not prime. The factors of a number are all integers that can be multiplied by another integer to produce the given number. For example, 2 is multiplied by 3 to produce 6. Therefore, 2 and 3 are both factors of 6. Similarly, $1 \times 6 = 6$ and $2 \times 3 = 6$, so 1, 2, 3, and 6 are all factors of 6. Another way to explain a factor is to say that a given number divides evenly by each of its factors to produce an integer. For example, 6 does not divide evenly by 5. Therefore, 5 is not a factor of 6.

Essentially, **factors** are the numbers multiplied to achieve a product. Thus, every product in a multiplication equation has, at minimum, two factors. Of course, some products will have more than two factors. For the sake of most discussions, assume that factors are positive integers.

To find a number's factors, start with 1 and the number itself. Then divide the number by 2, 3, 4, and so on, seeing if any divisors can divide the number without a remainder, keeping a list of those that do. Stop upon reaching either the number itself or another factor.

Let's find the factors of 45. Start with 1 and 45. Then try to divide 45 by 2, which fails. Now divide 45 by 3. The answer is 15, so 3 and 15 are now factors. Dividing by 4 doesn't work, and dividing by 5 leaves 9. Lastly, dividing 45 by 6, 7, and 8 all don't work. The next integer to try is 9, but this is already known to be a factor, so the factorization is complete. The factors of 45 are 1, 3, 5, 9, 15 and 45.

A **common factor** is a factor shared by two numbers. Let's take 45 and 30 and find the common factors:

> The factors of 45 are: 1, 3, 5, 9, 15, and 45.
> The factors of 30 are: 1, 2, 3, 5, 6, 10, 15, and 30.
> The common factors are 1, 3, 5, and 15.

The **greatest common factor** is the largest number among the shared, common factors. From the factors of 45 and 30, the common factors are 3, 5, and 15. Thus, 15 is the greatest common factor, as it's the largest number.

Multiples of a given number are found by taking that number and multiplying it by any other whole number. For example, 3 is a factor of 6, 9, and 12. Therefore, 6, 9, and 12 are multiples of 3. The multiples of any number are an infinite list. For example, the multiples of 5 are 5, 10, 15, 20, and so on. This list continues without end. A list of multiples is used in finding the **least common multiple**, or **LCM,** for fractions when a common denominator is needed. The denominators are written down and their multiples listed until a common number is found in both lists. This common number is the LCM.

If two numbers share no factors besides 1 in common, then their least common multiple will be simply their product. If two numbers have common factors, then their least common multiple will be their product divided by their greatest common factor. This can be visualized by the formula $LCM = \frac{x \times y}{GCF}$, where *x* and *y* are some integers, and LCM and GCF are their least common multiple and greatest common factor, respectively.

Prime factorization breaks down each factor of a whole number until only prime numbers remain. All composite numbers can be factored into prime numbers. For example, the prime factors of 12 are 2, 2, and 3 ($2 \times 2 \times 3 = 12$). To produce the prime factors of a number, the number is factored, and any composite numbers are continuously factored until the result is the product of prime factors only. A factor tree, such as the one below, is helpful when exploring this concept.

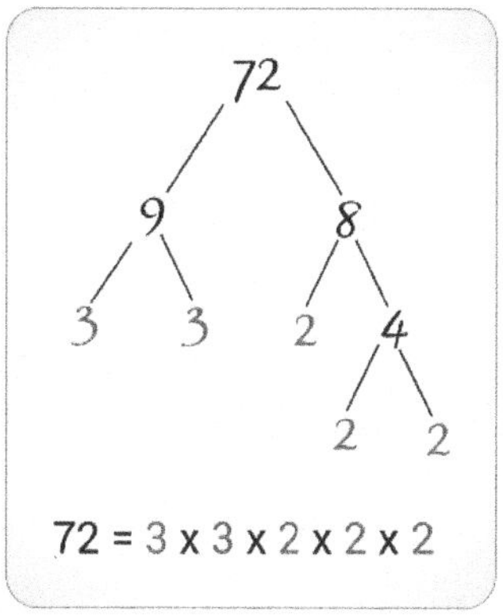

Let's break 129 down into its prime factors. First, the factors are 3 and 43. Both 3 and 43 are prime numbers, so we're done. But if 43 was not a prime number, then it would also need to be factorized until all of the factors are expressed as prime numbers.

Determining the Reasonableness of Results

The Reasonableness of Results

When solving math word problems, the solution obtained should make sense within the given scenario. The step of checking the solution will reduce the possibility of a calculation error or a solution that may be *mathematically* correct but not applicable in the real world. Consider the following scenarios:

A problem states that Lisa got 24 out of 32 questions correct on a test and asks to find the percentage of correct answers. To solve the problem, a student divided 32 by 24 to get 1.33, and then multiplied by 100 to get 133 percent. By examining the solution within the context of the problem, the student should recognize that getting all 32 questions correct will produce a perfect score of 100 percent. Therefore, a score of 133 percent with 8 incorrect answers does not make sense, and the calculations should be checked.

A problem states that the maximum weight on a bridge cannot exceed 22,000 pounds. The problem asks to find the maximum number of cars that can be on the bridge at one time if each car weighs 4,000 pounds. To solve this problem, a student divided 22,000 by 4,000 to get an answer of 5.5. By examining the solution within the context of the problem, the student should recognize that although the calculations are mathematically correct, the solution does not make sense. Half of a car on a bridge is not possible, so the student should determine that a maximum of 5 cars can be on the bridge at the same time.

Estimating

Estimation is finding a value that is close to a solution but is not the exact answer. For example, if there are values in the thousands to be multiplied, then each value can be estimated to the nearest thousand and the calculation performed. This value provides an approximate solution that can be determined very quickly.

Rounding is the process of either bumping a number up or down, based on a specified place value. First, the place value is specified. Then, the digit to its right is looked at. For example, if rounding to the nearest hundreds place, the digit in the tens place is used. If it is a 0, 1, 2, 3, or 4, the digit being rounded to is left alone. If it is a 5, 6, 7, 8 or 9, the digit being rounded to is increased by one. All other digits before the decimal point are then changed to zeros, and the digits in decimal places are dropped. If a decimal place is being rounded to, all subsequent digits are just dropped. For example, if 845,231.45 was to be rounded to the nearest thousands place, the answer would be 845,000. The 5 would remain the same due to the 2 in the hundreds place. Also, if 4.567 was to be rounded to the nearest tenths place, the answer would be 4.6. The 5 increased to 6 due to the 6 in the hundredths place, and the rest of the decimal is dropped.

Sometimes when performing operations such as multiplying numbers, the result can be estimated by rounding. For example, to estimate the value of 11.2×2.01, each number can be rounded to the nearest integer. This will yield a result of 22.

Rounding numbers helps with estimation because it changes the given number to a simpler, although less accurate, number than the exact given number. Rounding allows for easier calculations, which estimate the results of using the exact given number. The accuracy of the estimate and ease of use depends on the place value to which the number is rounded. Rounding numbers consists of:

- determining what place value the number is being rounded to
- examining the digit to the right of the desired place value to decide whether to round up or keep the digit, and
- replacing all digits to the right of the desired place value with zeros.

To round 746,311 to the nearest ten thousand, the digit in the ten thousands place should be located first. In this case, this digit is 4 (7<u>4</u>6,311). Then, the digit to its right is examined. If this digit is 5 or greater, the number will be rounded up by increasing the digit in the desired place by one. If the digit to the right of the place value being rounded is 4 or less, the number will be kept the same. For the given example, the digit being examined is a 6, which means that the number will be rounded up by increasing the digit to the left by one. Therefore, the digit 4 is changed to a 5. Finally, to write the rounded number, any digits to the left of the place value being rounded remain the same and any to its right are replaced with zeros. For the given example, rounding 746,311 to the nearest ten thousand will produce 750,000. To round 746,311 to the nearest hundred, the digit to the right of the three in the hundreds place is examined to determine whether to round up or keep the same number. In this case, that digit is a 1, so the number will be kept the same and any digits to its right will be replaced with zeros. The resulting rounded number is 746,300.

Rounding place values to the right of the decimal follows the same procedure, but digits being replaced by zeros can simply be dropped. To round 3.752891 to the nearest thousandth, the desired place value is located (3.75<u>2</u>891) and the digit to the right is examined. In this case, the digit 8 indicates that the number will be rounded up, and the 2 in the thousandths place will increase to a 3. Rounding up and replacing the digits to the right of the thousandths place produces 3.753000 which is equivalent to 3.753. Therefore, the zeros are not necessary, and the rounded number should be written as 3.753.

When rounding up, if the digit to be increased is a 9, the digit to its left is increased by 1 and the digit in the desired place value is changed to a zero. For example, the number 1,598 rounded to the nearest ten is 1,600. Another example shows the number 43.72961 rounded to the nearest thousandth is 43.730 or 43.73.

Mental math should always be considered as problems are worked through, and the ability to work through problems in one's head helps save time. If a problem is simple enough, such as $15 + 3 = 18$, it should be completed mentally. The ability to do this will increase once addition and subtraction in higher place values are grasped. Also, mental math is important in multiplication and division. The times tables multiplying all numbers from 1 to 12 should be memorized. This will allow for division within those numbers to be memorized as well. For example, we should know easily that $121 \div 11 = 11$ because it should be memorized that $11 \times 11 = 121$. Here is the multiplication table to be memorized:

x	1	2	3	4	5	6	7	8	9	10	11	12	13	14	15
1	1	2	3	4	5	6	7	8	9	10	11	12	13	14	15
2	2	4	6	8	10	12	14	16	18	20	22	24	26	28	30
3	3	6	9	12	15	18	21	24	27	30	33	36	39	42	45
4	4	8	12	16	20	24	28	32	36	40	44	48	52	56	60
5	5	10	15	20	25	30	35	40	45	50	55	60	65	70	75
6	6	12	18	24	30	36	42	48	54	60	66	72	78	84	90
7	7	14	21	28	35	42	49	56	63	70	77	84	91	98	105
8	8	16	24	32	40	48	56	64	72	80	88	96	104	112	120
9	9	18	27	36	45	54	63	72	81	90	99	108	117	126	135
10	10	20	30	40	50	60	70	80	90	100	110	120	130	140	150
11	11	22	33	44	55	66	77	88	99	110	121	132	143	154	165
12	12	24	36	48	60	72	84	96	108	120	132	144	156	168	180
13	13	26	39	52	65	78	91	104	117	130	143	156	169	182	195
14	14	28	42	56	70	84	98	112	126	140	154	168	182	196	210
15	15	30	45	60	75	90	105	120	135	150	165	180	195	210	225

The values along the diagonal of the table consist of **perfect squares**. A perfect square is a number that represents a product of two equal integers.

Estimating Absolute and Relative Error in the Numerical Answer to a Problem

Once a result is determined to be logical within the context of a given problem, the result should be evaluated by its nearness to the expected answer. This is performed by approximating given values to perform mental math. Numbers should be rounded to the nearest value possible to check the initial results.

Consider the following example: A problem states that a customer is buying a new sound system for their home. The customer purchases a stereo for $435, 2 speakers for $67 each, and the necessary cables for $12. The customer chooses an option that allows him to spread the costs over equal payments for 4 months. How much will the monthly payments be?

After making calculations for the problem, a student determines that the monthly payment will be $145.25. To check the accuracy of the results, the student rounds each cost to the nearest ten ($440 + 70 + 70 + 10$) and determines that the total is approximately $590. Dividing by 4 months gives an approximate monthly payment of $147.50. Therefore, the student can conclude that the solution of $145.25 is very close to what should be expected.

When rounding, the place-value that is used in rounding can make a difference. Suppose the student had rounded to the nearest hundred for the estimation. The result ($400 + 100 + 100 + 0 = 600;\ 600 \div 4 = 150$) will show that the answer is reasonable but not as close to the actual value as rounding to the nearest ten.

When considering the accuracy of estimates, the error in the estimated solution can be shown as absolute and relative. **Absolute error** tells the actual difference between the estimated value and the true, calculated value. The **relative error** tells how large the error is in relation to the true value. There may be two problems where the absolute error of the values (the estimated one and the calculated one) is 10. For one problem, this may mean the relative error in the estimate is very small because the estimated value is 15,000, and the true value is 14,990. Ten in relation to the true value of 15,000 is small: 0.06%. For the other problem, the estimated value is 50, and the true value is 40. In this case, the absolute error of 10 means a high relative error because the true value is smaller. The relative error is 10/40 = 0.25 or 25%.

Algebra

Algebraic Expressions, Equations, and Formulas

Performing Arithmetic Operations on Polynomials

An expression of the form ax^n, where *n* is a non-negative integer, is called a **monomial** because it contains one term. A sum of monomials is called a **polynomial**. For example, $-4x^3 + x$ is a polynomial, while $5x^7$ is a monomial. A function equal to a polynomial is called a **polynomial function**.

The monomials in a polynomial are also called the **terms** of the polynomial.

The constants that precede the variables are called **coefficients**.

The highest value of the exponent of *x* in a polynomial is called the **degree** of the polynomial. So, $-4x^3 + x$ has a degree of 3, while $-2x^5 + x^3 + 4x + 1$ has a degree of 5. When multiplying polynomials, the degree of the result will be the sum of the degrees of the two polynomials being multiplied.

Addition and subtraction operations can be performed on polynomials with like terms. **Like terms** refers to terms that have the same variable and exponent. The two following polynomials can be added together by collecting like terms:

$$(x^2 + 3x - 4) + (4x^2 - 7x + 8)$$

The x^2 terms can be added as $x^2 + 4x^2 = 5x^2$. The x terms can be added as $3x + -7x = -4x$, and the constants can be added as $-4 + 8 = 4$. The following expression is the result of the addition:

$$5x^2 - 4x + 4$$

When subtracting polynomials, the same steps are followed, only subtracting like terms together.

Multiplication of polynomials can also be performed. Given the two polynomials, $(y^3 - 4)$ and $(x^2 + 8x - 7)$, each term in the first polynomial must be multiplied by each term in the second polynomial. The steps to multiply each term in the given example are as follows:

$$(y^3 \times x^2) + (y^3 \times 8x) + (y^3 \times -7) + (-4 \times x^2) + (-4 \times 8x) + (-4 \times -7)$$

Simplifying each multiplied part, yields:

$$x^2y^3 + 8xy^3 - 7y^3 - 4x^2 - 32x + 28$$

None of the terms can be combined because there are no like terms in the final expression. Any polynomials can be multiplied by each other by following the same set of steps, then collecting like terms at the end.

Manipulating and Performing Arithmetic Operations on Rational Expressions

A fraction, or ratio, wherein each part is a polynomial, defines **rational expressions**. Some examples include $\frac{2x+6}{x}$, $\frac{1}{x^2-4x+8}$, and $\frac{z^2}{x+5}$. Exponents on the variables are restricted to whole numbers, which means roots and negative exponents are not included in rational expressions.

Rational expressions can be transformed by factoring. For example, the expression $\frac{x^2-5x+6}{(x-3)}$ can be rewritten by factoring the numerator to obtain $\frac{(x-3)(x-2)}{(x-3)}$. Therefore, the common binomial $(x - 3)$ can cancel so that the simplified expression is $\frac{(x-2)}{1} = (x - 2)$.

Additionally, other rational expressions can be rewritten to take on different forms. Some may be factorable in themselves, while others can be transformed through arithmetic operations. Rational expressions are closed under addition, subtraction, multiplication, and division by a nonzero expression. **Closed** means that if any one of these operations is performed on a rational expression, the result will still be a rational expression. The set of all real numbers is another example of a set closed under all four operations.

Adding and subtracting rational expressions is based on the same concepts as adding and subtracting simple fractions. For both concepts, the denominators must be the same for the operation to take place. For example, here are two rational expressions:

$$\frac{x^3 - 4}{(x - 3)} + \frac{x + 8}{(x - 3)}$$

Since the denominators are both $(x - 3)$, the numerators can be combined by collecting like terms to form:

$$\frac{x^3 + x + 4}{(x - 3)}$$

If the denominators are different, they need to be made common (the same) by using the **Least Common Denominator (LCD)**. Each denominator needs to be factored, and the LCD contains each factor that appears in any one denominator the greatest number of times it appears in any denominator. The original expressions need to be multiplied by a form of 1 such as 5/5 or x-2/x-2, which will turn each denominator into the LCD. This process is like adding fractions with unlike denominators. It is also important when working with rational expressions to define what value of the variable makes the denominator zero. For this particular value, the expression is undefined.

Consider this problem as an example of using rational expressions. Reggie wants to lay sod in his rectangular backyard. The length of the yard is given by the expression $4x + 2$ and the width is unknown. The area of the yard is $20x + 10$. Reggie needs to find the width of the yard. Knowing that the area of a rectangle is length multiplied by width, an expression can be written to find the width: $\frac{20x+10}{4x+2}$, area divided by length. Simplifying this expression by factoring out 10 on the top and 2 on the bottom leads to this expression: $\frac{10(2x+1)}{2(2x+1)}$. By cancelling out the $2x + 1$, that results in $\frac{10}{2} = 5$. The width of the yard is found to be 5 by simplifying a rational expression.

Evaluating, Manipulating, and Comparing Algebraic Expressions Involving Radicals and Exponents

Here are some of the most important properties of exponents and roots: if *n* is an integer, and if $a^n = b^n$, then $a = b$ if *n* is odd; but $a = \pm b$ if *n* is even. Similarly, if the roots of two things are equal, $\sqrt[n]{a} = \sqrt[n]{b}$, then $a = b$. This means that when starting with a true equation, both sides of that equation can be raised to a given power to obtain another true equation. Beware that when an even-powered root is taken on both sides of the equation, a $\pm$ in the result. For example, given the equation $x^2 = 16$, take the square root of both sides to solve for x. This results in the answer $x = \pm 4$ because $(-4)^2 = 16$ and $(4)^2 = 16$.

Another property is that if $a^n = a^m$, then $n = m$. This is true for any real numbers *n* and *m*.

For solving the equation $\sqrt{x+2} - 1 = 3$, start by moving the -1 over to the right-hand side. This is performed by adding 1 to both sides, which yields $\sqrt{x+2} = 4$. Now, square both sides, but remember that by squaring both sides, the signs are irrelevant. This yields $x + 2 = 16$, which simplifies to give $x = 14$.

Now consider the problem $(x + 1)^4 = 16$. To solve this, take the 4th root of both sides, which means an ambiguity in the sign will be introduced because it is an even root: $\sqrt[4]{(x+1)^4} = \pm\sqrt[4]{16}$. The right-hand side is 2, since $2^4 = 16$. Therefore, $x + 1 = \pm 2$ or $x = -1 \pm 2$. Thus, the two possible solutions are $x = -3$ and $x = 1$.

Remember that when solving equations, the answer can be checked by plugging the solution back into the problem to make a true statement.

Using Variables to Construct and Solve Equations in Real-World Contexts

An algebraic expression is a statement about an unknown quantity expressed in mathematical symbols. A variable is used to represent the unknown quantity, usually denoted by a letter. An equation is a statement in which two expressions (at least one containing a variable) are equal to each other. An algebraic expression can be thought of as a mathematical phrase and an equation can be thought of as a mathematical sentence.

Algebraic expressions and equations both contain numbers, variables, and mathematical operations. The following are examples of algebraic expressions: $5x + 3, 7xy - 8(x^2 + y)$, and $\sqrt{a^2 + b^2}$. An expression can be simplified or evaluated for given values of variables. The following are examples of equations: $2x + 3 = 7$, $a^2 + b^2 = c^2$, and $2x + 5 = 3x - 2$. An equation contains two sides separated by an equal sign. Equations can be solved to determine the value(s) of the variable for which the statement is true.

An algebraic expression is a statement about unknown quantities expressed in mathematical symbols. The statement *five times a number added to forty* is expressed as $5x + 40$. An equation is a statement in which two expressions (with at least one containing a variable) are equal to one another. The statement *five times a number added to forty is equal to ten* is expressed as $5x + 40 = 10$.

Real world scenarios can also be expressed mathematically. Suppose a job pays its employees $300 per week and $40 for each sale made. The weekly pay is represented by the expression $40x + 300$ where *x* is the number of sales made during the week.

Consider the following scenario: Bob had $20 and Tom had $4. After selling 4 ice cream cones to Bob, Tom has as much money as Bob. The cost of an ice cream cone is an unknown quantity and can be represented by a variable (*x*). The amount of money Bob has after his purchase is four times the cost of an ice cream cone subtracted from his original $\$20 \rightarrow 20 - 4x$. The amount of money Tom has after his sale is four times the cost of an ice cream cone added to his original $\$4 \rightarrow 4x + 4$. After the sale, the amount of money that Bob and Tom have are equal $\rightarrow 20 - 4x = 4x + 4$.

Translating Verbal Relationships into Algebraic Equations or Expressions

When expressing a verbal or written statement mathematically, it is vital to understand words or phrases that can be represented with symbols. The following are examples:

Symbol	Phrase
+	Added to; increased by; sum of; more than
−	Decreased by; difference between; less than; take away
×	Multiplied by; 3(4,5...) times as large; product of
÷	Divided by; quotient of; half (third, etc.) of
=	Is; the same as; results in; as much as; equal to
x,t,n, etc.	A number; unknown quantity; value of; variable

Addition and subtraction are **inverse operations**. Adding a number and then subtracting the same number will cancel each other out, resulting in the original number, and vice versa. For example, $8 + 7 - 7 = 8$ and $137 - 100 + 100 = 137$. Similarly, multiplication and division are inverse operations. Therefore, multiplying by a number and then dividing by the same number results in the original number, and vice versa. For example, $8 \times 2 \div 2 = 8$ and $12 \div 4 \times 4 = 12$. Inverse operations are used to work backwards to solve problems. In the case that 7 and a number add to 18, the inverse operation of subtraction is used to find the unknown value ($18 - 7 = 11$). If a school's entire 4th grade was divided evenly into 3 classes each with 22 students, the inverse operation of multiplication is used to determine the total students in the grade ($22 \times 3 = 66$). Additional scenarios involving inverse operations are included in the tables below.

There are a variety of real-world situations in which one or more of the operators is used to solve a problem. The tables below display the most common scenarios.

Addition & Subtraction

	Unknown Result	Unknown Change	Unknown Start
Adding to	5 students were in class. 4 more students arrived. How many students are in class? $5 + 4 = ?$	8 students were in class. More students arrived late. There are now 18 students in class. How many students arrived late? $8 + ? = 18$ Solved by inverse operations $18 - 8 = ?$	Some students were in class early. 11 more students arrived. There are now 17 students in class. How many students were in class early? $? + 11 = 17$ Solved by inverse operations $17 - 11 = ?$
Taking from	15 students were in class. 5 students left class. How many students are in class now? $15 - 5 = ?$	12 students were in class. Some students left class. There are now 8 students in class. How many students left class? $12 - ? = 8$ Solved by inverse operations $8 + ? = 12 \rightarrow 12 - 8 = ?$	Some students were in class. 3 students left class. Then there were 13 students in class. How many students were in class before? $? - 3 = 13$ Solved by inverse operations $13 + 3 = ?$

	Unknown Total	Unknown Addends (Both)	Unknown Addends (One)
Putting together/ taking apart	The homework assignment is 10 addition problems and 8 subtraction problems. How many problems are in the homework assignment? $10 + 8 = ?$	Bobby has $9. How much can Bobby spend on candy and how much can Bobby spend on toys? $9 = ? + ?$	Bobby has 12 pairs of pants. 5 pairs of pants are shorts, and the rest are long. How many pairs of long pants does he have? $12 = 5 + ?$ Solved by inverse operations $12 - 5 = ?$

	Unknown Difference	**Unknown Larger Value**	**Unknown Smaller Value**
Comparing	Bobby has 5 toys. Tommy has 8 toys. How many more toys does Tommy have than Bobby? $5+?=8$ Solved by inverse operations $8-5=?$ Bobby has \$6. Tommy has \$10. How many fewer dollars does Bobby have than Tommy? $10-6=?$	Tommy has 2 more toys than Bobby. Bobby has 4 toys. How many toys does Tommy have? $2+4=?$ Bobby has 3 fewer dollars than Tommy. Bobby has \$8. How many dollars does Tommy have? $?-3=8$ Solved by inverse operations $8+3=?$	Tommy has 6 more toys than Bobby. Tommy has 10 toys. How many toys does Bobby have? $?+6=10$ Solved by inverse operations $10-6=?$ Bobby has \$5 less than Tommy. Tommy has \$9. How many dollars does Bobby have? $9-5=?$

Multiplication and Division

	Unknown Product	**Unknown Group Size**	**Unknown Number of Groups**
Equal groups	There are 5 students, and each student has 4 pieces of candy. How many pieces of candy are there in all? $5\times 4=?$	14 pieces of candy are shared equally by 7 students. How many pieces of candy does each student have? $7\times ?=14$ Solved by inverse operations $14\div 7=?$	If 18 pieces of candy are to be given out 3 to each student, how many students will get candy? $?\times 3=18$ Solved by inverse operations $18\div 3=?$

	Unknown Product	Unknown Factor	Unknown Factor
Arrays	There are 5 rows of students with 3 students in each row. How many students are there? $5 \times 3 = ?$	If 16 students are arranged into 4 equal rows, how many students will be in each row? $4 \times ? = 16$ Solved by inverse operations $16 \div 4 = ?$	If 24 students are arranged into an array with 6 columns, how many rows are there? $? \times 6 = 24$ Solved by inverse operations $24 \div 6 = ?$

	Larger Unknown	Smaller Unknown	Multiplier Unknown
Comparing	A small popcorn costs \$1.50. A large popcorn costs 3 times as much as a small popcorn. How much does a large popcorn cost? $1.50 \times 3 = ?$	A large soda costs \$6 and that is 2 times as much as a small soda costs. How much does a small soda cost? $2 \times ? = 6$ Solved by inverse operations $6 \div 2 = ?$	A large pretzel costs \$3 and a small pretzel costs \$2. How many times as much does the large pretzel cost as the small pretzel? $? \times 2 = 3$ Solved by inverse operations $3 \div 2 = ?$

Recognizing and Representing Linear Relationships Algebraically

Linear relationships describe the way two quantities change with respect to each other. The relationship is defined as linear because a line is produced if all the sets of corresponding values are graphed on a coordinate grid. When expressing the linear relationship as an equation, the equation is often written in the form $y = mx + b$ (slope-intercept form) where *m* and *b* are numerical values and *x* and *y* are variables (for example, $y = 5x + 10$). Given a linear equation and the value of either variable (*x* or *y*), the value of the other variable can be determined.

Suppose a teacher is grading a test containing 20 questions with 5 points given for each correct answer, adding a curve of 10 points to each test. This linear relationship can be expressed as the equation $y = 5x + 10$ where *x* represents the number of correct answers, and y represents the test score. To determine the score of a test with a given number of correct answers, the number of correct answers is substituted into the equation for *x* and evaluated. For example, for 10 correct answers, 10 is substituted for *x*: $y = 5(10) + 10 \rightarrow y = 60$. Therefore, 10 correct answers will result in a score of 60. The number of correct answers needed to obtain a certain score can also be determined. To determine the number of correct answers needed to score a 90, 90 is substituted for *y* in the equation (*y* represents the test score) and solved: $90 = 5x + 10 \rightarrow 80 = 5x \rightarrow 16 = x$. Therefore, 16 correct answers are needed to score a 90.

Linear relationships may be represented by a table of 2 corresponding values. Certain tables may determine the relationship between the values and predict other corresponding sets. Consider the table below, which displays the money in a checking account that charges a monthly fee:

Month	0	1	2	3	4
Balance	$210	$195	$180	$165	$150

An examination of the values reveals that the account loses $15 every month (the month increases by one and the balance decreases by 15). This information can be used to predict future values. To determine what the value will be in month 6, the pattern can be continued, and it can be concluded that the balance will be $120. To determine which month the balance will be $0, $210 is divided by $15 (since the balance decreases $15 every month), resulting in month 14.

Similar to a table, a graph can display corresponding values of a linear relationship.

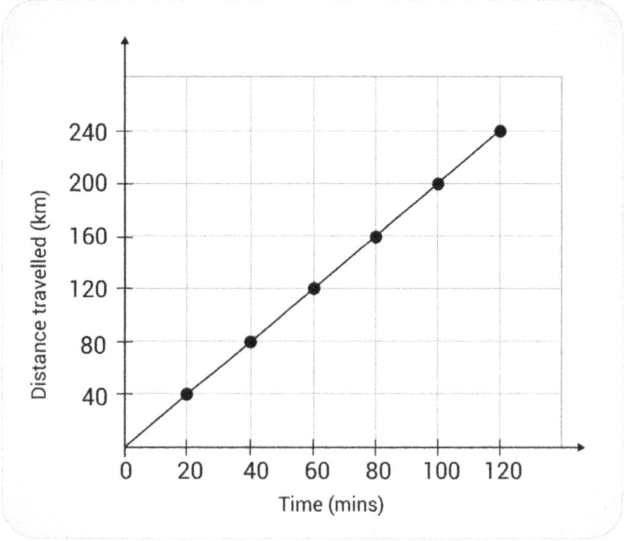

The graph above represents the relationship between distance traveled and time. To find the distance traveled in 80 minutes, the mark for 80 minutes is located at the bottom of the graph. By following this mark directly up on the graph, the corresponding point for 80 minutes is directly across from the 160 kilometer mark. This information indicates that the distance travelled in 80 minutes is 160 kilometers. To predict information not displayed on the graph, the way in which the variables change with respect to one another is determined. In this case, distance increases by 40 kilometers as time increases by 20 minutes. This information can be used to continue the data in the graph or convert the values to a table.

Determining the Equation of a Line

A function is called **linear** if it can take the form of the equation $f(x) = ax + b$, or $y = ax + b$, for any two numbers *a* and *b*. A linear equation forms a straight line when graphed on the coordinate plane. An example of a linear function is shown below on the graph.

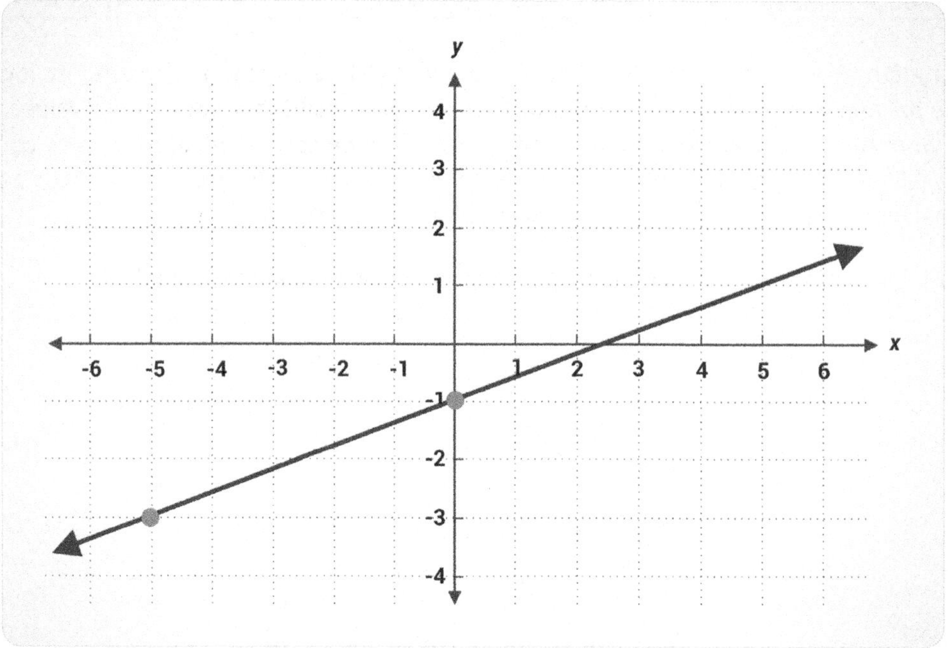

This is a graph of the following function: $y = \frac{2}{5}x - 1$. A table of values that satisfies this function is shown below.

x	y
-5	-3
0	-1
5	1
10	3

These points can be found on the graph using the form (x,y).

Given the graph of a line, its equation can be written in two ways. If the *y*-intercept is easily identified (is an integer), it and another point can be used to determine the slope. When determining $\frac{change\ in\ y}{change\ in\ x}$ from one point to another on the graph, the distance for $\frac{rise}{run}$ is being figured. The equation should be written in slope-intercept form, $y = mx + b$, with *m* representing the slope and *b* representing the *y*-intercept.

The equation of a line can also be written by identifying two points on the graph of the line. To do so, the slope is calculated and then the values are substituted for the slope and either of the ordered pairs into the point-slope form of an equation.

The Basic Forms of Linear Equations

Linear functions may take forms other than $y = ax + b$. The most common forms of linear equations are explained below:

1. Standard Form: $Ax + By = C$, in which the slope is given by $m = \frac{-A}{B}$, and the *y*-intercept is given by $\frac{C}{B}$.

2. Slope-Intercept Form: $y = mx + b$, where the slope is *m* and the *y* intercept is *b*.

3. Point-Slope Form: $y - y_1 = m(x - x_1)$, where the slope is *m* and (x_1, y_1) is any point on the chosen line.

4. Two-Point Form: $\frac{y - y_1}{x - x_1} = \frac{y_2 - y_1}{x_2 - x_1}$, where (x_1, y_1) and (x_2, y_2) are any two distinct points on the chosen line. Note that the slope is given by $m = \frac{y_2 - y_1}{x_2 - x_1}$.

5. Intercept Form: $\frac{x}{x_1} + \frac{y}{y_1} = 1$, in which x_1 is the *x*-intercept and y_1 is the *y*-intercept.

These five ways to write linear equations are all useful in different circumstances. Depending on the given information, it may be easier to write one of the forms over another.

If $y = mx$, *y* is directly proportional to *x*. In this case, changing *x* by a factor changes *y* by that same factor. If $y = \frac{m}{x}$, *y* is inversely proportional to *x*. For example, if *x* is increased by a factor of 3, then *y* will be decreased by the same factor, 3.

Solving Linear Equations and Inequalities

Solving One-Variable Linear Equations and Inequalities Algebraically and Represent Solutions on a Number Line

Linear equations and linear inequalities are both comparisons of two algebraic expressions. However, unlike equations in which the expressions are equal, linear inequalities compare expressions that may be unequal. Linear equations typically have one value for the variable that makes the statement true. Linear inequalities generally have an infinite number of values that make the statement true.

When solving a linear equation, the desired result requires determining a numerical value for the unknown variable. If given a linear equation involving addition, subtraction, multiplication, or division, working backwards isolates the variable. Addition and subtraction are inverse operations, as are multiplication and division. Therefore, they can be used to cancel each other out.

The first steps to solving linear equations are distributing, if necessary, and combining any like terms on the same side of the equation. Sides of an equation are separated by an **equal sign**. Next, the equation is manipulated to show the variable on one side. Whatever is done to one side of the equation must be done to the other side of the equation to remain equal. Inverse operations are then used to isolate the variable and undo the order of operations backwards. Addition and subtraction are undone, then multiplication and division are undone.

For example, solve $4(t - 2) + 2t - 4 = 2(9 - 2t)$

Distributing: $4t - 8 + 2t - 4 = 18 - 4t$

Combining like terms: $6t - 12 = 18 - 4t$

Adding $4t$ to each side to move the variable: $10t - 12 = 18$

Adding 12 to each side to isolate the variable: $10t = 30$

Dividing each side by 10 to isolate the variable: $t = 3$

The answer can be checked by substituting the value for the variable into the original equation, ensuring that both sides calculate to be equal.

Linear inequalities express the relationship between unequal values. More specifically, they describe in what way the values are unequal. A value can be greater than (>), less than (<), greater than or equal to (≥), or less than or equal to (≤) another value. $5x + 40 > 65$ is read as *five times a number added to forty is greater than sixty-five.*

When solving a linear inequality, the solution is the set of all numbers that make the statement true. The inequality $x + 2 \geq 6$ has a solution set of 4 and every number greater than 4 (4.01; 5; 12; 107; etc.). Adding 2 to 4 or any number greater than 4 results in a value that is greater than or equal to 6. Therefore, $x \geq 4$ is the solution set.

To algebraically solve a linear inequality, follow the same steps as those for solving a linear equation. The inequality symbol stays the same for all operations *except* when multiplying or dividing by a negative number. If multiplying or dividing by a negative number while solving an inequality, the relationship reverses (the sign flips). In other words, > switches to < and vice versa. Multiplying or dividing by a positive number does not change the relationship, so the sign stays the same. An example is shown below.

Solve $-2x - 8 \leq 22$

Add 8 to both sides: $-2x \leq 30$

Divide both sides by -2: $x \geq -15$

Solutions of a linear equation or a linear inequality are the values of the variable that make a statement true. In the case of a linear equation, the solution set (list of all possible solutions) typically consists of a single numerical value. To find the solution, the equation is solved by isolating the variable. For example, solving the equation $3x - 7 = -13$ produces the solution $x = -2$. The only value for *x* which produces a true statement is -2. This can be checked by substituting -2 into the original equation to check that both sides are equal. In this case, $3(-2) - 7 = -13 \rightarrow -13 = -13$; therefore, -2 is a solution.

Although linear equations generally have one solution, this is not always the case. If there is no value for the variable that makes the statement true, there is no solution to the equation. Consider the equation $x + 3 = x - 1$. There is no value for *x* in which adding 3 to the value produces the same result as subtracting one from the value. Conversely, if any value for the variable makes a true statement, the equation has an infinite number of solutions. Consider the equation $3x + 6 = 3(x + 2)$. Any number substituted for *x* will result in a true statement (both sides of the equation are equal).

By manipulating equations like the two above, the variable of the equation will cancel out completely. If the remaining constants express a true statement (ex. $6 = 6$), then all real numbers are solutions to the equation. If the constants left express a false statement (ex. $3 = -1$), then no solution exists for the equation.

Solving a linear inequality requires all values that make the statement true to be determined. For example, solving $3x - 7 \geq -13$ produces the solution $x \geq -2$. This means that -2 and any number greater than -2 produces a true statement. Solution sets for linear inequalities will often be displayed using a number line. If a value is included in the set (≥ or ≤), a shaded dot is placed on that value and an arrow extending in the direction of the solutions. For a variable > or ≥ a number, the arrow will point right on a number line, the direction where the numbers increase. If a variable is < or ≤ a number, the arrow will point left on a number line, which is the direction where the numbers decrease. If the value is not included in the set (> or <), an open (unshaded) circle on that value is used with an arrow in the appropriate direction.

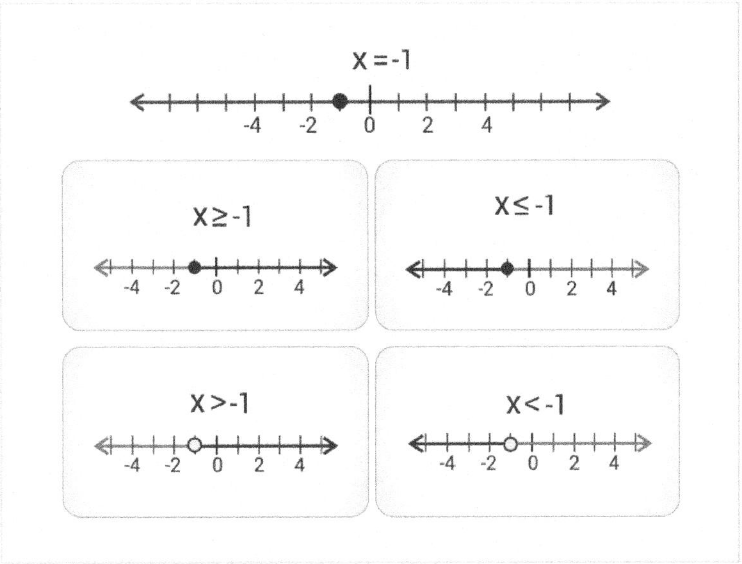

Similar to linear equations, a linear inequality may have a solution set consisting of all real numbers or can contain no solution. When solved algebraically, a linear inequality in which the variable cancels out and results in a true statement (ex. $7 \geq 2$) has a solution set of all real numbers. A linear inequality in which the variable cancels out and results in a false statement (ex. $7 \leq 2$) has no solution.

Imagine the following problem: The sum of a number and 5 is equal to -8 times the number.

To find this unknown number, a simple equation can be written to represent the problem. Key words such as difference, equal, and times are used to form the following equation with one variable: $n + 5 = -8n$. When solving for n, opposite operations are used. First, n is subtracted from $-8n$ across the equals sign, resulting in $5 = -9n$. Then, -9 is divided on both sides, leaving $n = -\frac{5}{9}$. This solution can be graphed on the number line with a dot as shown below:

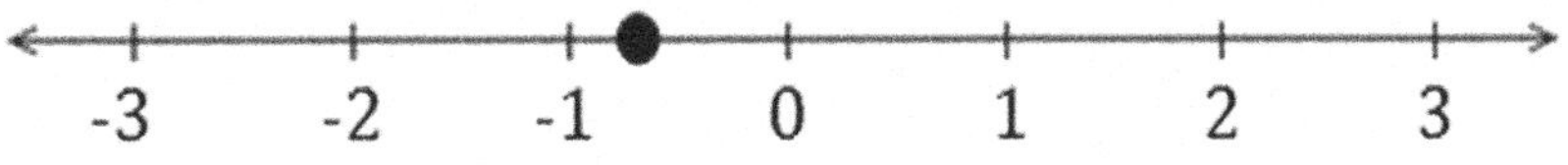

If the problem were changed to say, "The sum of a number and 5 is greater than -8 times the number," then an inequality would be used instead of an equation. Using key words again, *greater than* is represented by the symbol >. The inequality $n + 5 > -8n$ can be solved using the same techniques, resulting in $n < -\frac{5}{9}$. The only time solving an inequality differs from solving an equation is when a negative number is either multiplied by or divided by each side of the inequality. The sign must be switched in this case. For this example, the graph of the solution changes to the following graph because the solution represents all real numbers less than $-\frac{5}{9}$. Not included in this solution is $-\frac{5}{9}$ because it is a *less than* symbol, not *equal to*.

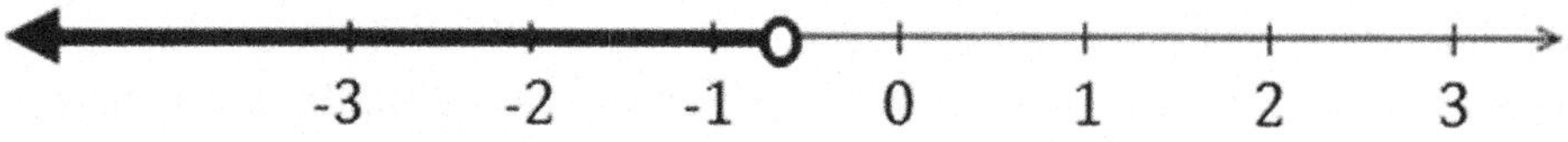

A graph of the solution set for a linear inequality shows the ordered pairs that make the statement true. The graph consists of a boundary line dividing the coordinate plane and shading on one side of the boundary. The boundary line should be graphed just as a linear equation would be graphed. If the inequality symbol is $>$ or $<$, a dashed line can be used to indicate that the line is not part of the solution set. If the inequality symbol is $\geq$ or $\leq$, a solid line can be used to indicate that the boundary line is included in the solution set. An ordered pair (*x*, *y*) on either side of the line should be chosen to test in the inequality statement. If substituting the values for *x* and *y* results in a true statement ($15(3) + 25(2) > 90$), that ordered pair and all others on that side of the boundary line are part of the solution set. To indicate this, that region of the graph should be shaded. If substituting the ordered pair results in a false statement, the ordered pair and all others on that side are not part of the solution set.

Therefore, the other region of the graph contains the solutions and should be shaded.

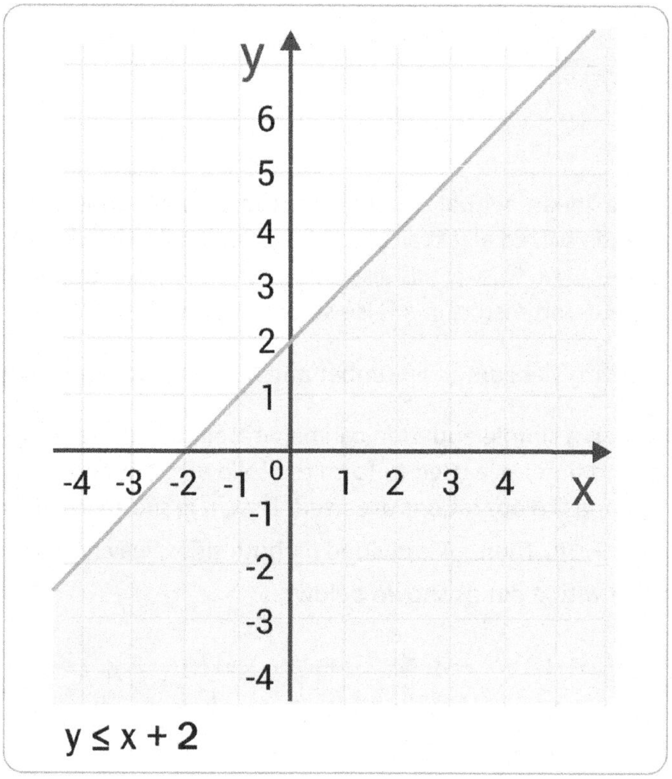

A question may simply ask whether a given ordered pair is a solution to a given inequality. To determine this, the values should be substituted for the ordered pair into the inequality. If the result is a true statement, the ordered pair is a solution; if the result is a false statement, the ordered pair is not a solution.

Solving Equations

Solving equations in one variable is the process of isolating that variable on one side of the equation. The letters in an equation are variables as they stand for unknown quantities that you are trying to solve for. The numbers attached to the variables by multiplication are called coefficients. *X* is commonly used as a variable, though any letter can be used. For example, in $3x - 7 = 20$, the variable is $3x$, and it needs to be isolated. The numbers (also called **constants**) are -7 and 20. That means $3x$ needs to be on one side of the equals sign (either side is fine), and all the numbers need to be on the other side of the equals sign.

To accomplish this, the equation must be manipulated by performing opposite operations of what already exists. Remember that addition and subtraction are opposites and that multiplication and division are opposites. Any action taken to one side of the equation must be taken on the other side to maintain equality.

So, since the 7 is being subtracted, it can be moved to the right side of the equation by adding seven to both sides:

$$3x - 7 = 20$$

$$3x - 7 + 7 = 20 + 7$$

$$3x = 27$$

Now that the variable $3x$ is on one side and the constants (now combined into one constant) are on the other side, the 3 needs to be moved to the right side. 3 and x are being multiplied together, so 3 then needs to be divided from each side.

$$\frac{3x}{3} = \frac{27}{3}$$

$$x = 9$$

Now that x has been completely isolated, we know its value.

The solution is found to be $x = 9$. This solution can be checked for accuracy by plugging $x = 9$ in the original equation. After simplifying the equation, $20 = 20$ is found, which is a true statement:

$$3 \times 9 - 7 = 20$$

$$27 - 7 = 20$$

$$20 = 20$$

Equations that require solving for a variable (**algebraic equations**) come in many forms. Here are some more examples:

Sometimes, no number is attached to the variable:

$$x + 8 = 20$$

$$x + 8 - 8 = 20 - 8$$

$$x = 12$$

A fraction may be in the variable:

$$\frac{1}{2}z + 24 = 36$$

$$\frac{1}{2}z + 24 - 24 = 36 - 24$$

$$\frac{1}{2}z = 12$$

Now we multiply the fraction by its inverse:

$$\frac{2}{1} \times \frac{1}{2}z = 12 \times \frac{2}{1}$$

$$z = 24$$

Some algebraic equations contain multiple instances of x:

$$14x + x - 4 = 3x + 2$$

All instances of x can be combined.

$$15x - 4 = 3x + 2$$

$$15x - 4 + 4 = 3x + 2 + 4$$

$$15x = 3x + 6$$

$$15x - 3x = 3x + 6 - 3x$$

$$12x = 6$$

$$\frac{12x}{12} = \frac{6}{12}$$

$$x = \frac{1}{2}$$

Nonlinear Equations and Inequalities

Solving One-Variable Nonlinear Equations and Inequalities Algebraically and Represent Solutions on a Number Line

Not all functional relationships are linear, and some **nonlinear functions** exhibit behavior that is helpful when modeling real-world situations. Exponential and trigonometric functions are specific nonlinear functions that are widely used. If a quantity decays or grows at changing rate, exponential functions are used. Common applications involve population growth and decay. Real-world scenarios that involve periodic motion, such as waves or pendulums, can be modeled using trigonometric functions due to their periodic nature. Quadratic and exponential functions are two of the more common nonliner functions that will be encountered.

Quadratic Functions

A **quadratic function** can be written in the form $y = ax^2 + bx + c$. The u-shaped graph of a quadratic function is called a **parabola**. The graph can either open up or open down (upside down u). The graph is symmetric about a vertical line, called the **axis of symmetry**. Corresponding points on the parabola are directly across from each other (same *y*-value) and are the same distance from the axis of symmetry (on either side). The axis of symmetry intersects the parabola at its **vertex**. The *y*-value of the vertex represents the minimum or maximum value of the function. If the graph opens up, the value of *a* in its equation is positive, and the vertex represents the minimum of the function. If the graph opens down, the value of *a* in its equation is negative, and the vertex represents the maximum of the function.

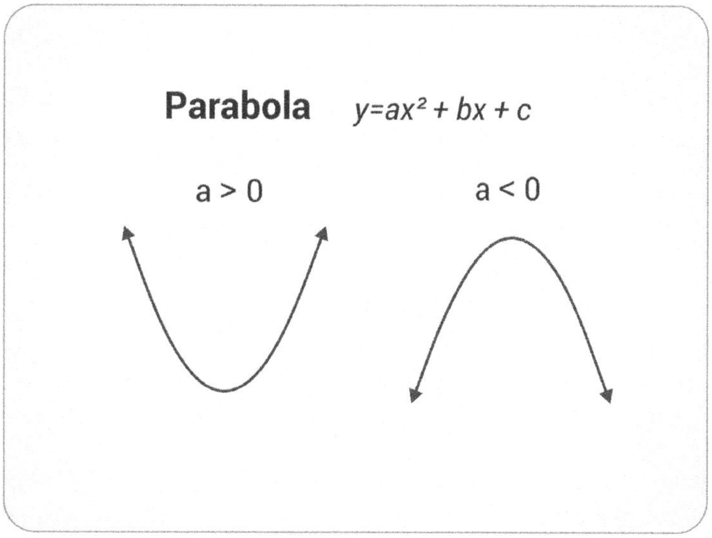

A quadratic function can be written in the standard form: $y = ax^2 + bx + c$. It can be represented by a u-shaped graph called a parabola. For a quadratic function where the value of *a* is positive, as the inputs increase, the outputs increase until a certain value (maximum of the function) is reached. As inputs increase past the value that corresponds with the maximum output, the relationship reverses, and the outputs decrease. For a quadratic function where *a* is negative, as the inputs increase, the outputs (1) decrease, (2) reach a maximum, and (3) then increase.

Consider a ball thrown straight up into the air. As time passes, the height of the ball increases until it reaches its maximum height. After reaching the maximum height, as time increases, the height of the ball decreases (it is falling toward the ground). This relationship can be expressed as a quadratic function where time is the input (*x*), and the height of the ball is the output (*y*).

Equations with one variable (linear equations) can be solved using the addition principle and multiplication principle. If $a = b$, then $a + c = b + c$, and $ac = bc$. Given the equation $2x - 3 = 5x + 7$, the first step is to combine the variable terms and the constant terms. Using the principles, expressions can be added and subtracted onto and off both sides of the equals sign, so the equation turns into $-10 = 3x$. Dividing by 3 on both sides through the multiplication principle with $c = \frac{1}{3}$ results in the final answer of $x = \frac{-10}{3}$.

However, this same process cannot be used to solve nonlinear equations, including quadratic equations. Quadratic equations have a higher degree than linear ones (2 versus 1) and are not solved by simply using opposite operations. When an equation has a degree of 2, completing the square is an option. For example, the quadratic equation $x^2 - 6x + 2 = 0$ can be rewritten by completing the square. The goal of completing the square is to get the equation into the form $(x - p)^2 = q$. Using the example, the constant term 2 first needs to be moved over to the opposite side by subtracting. Then, the square can be completed by adding 9 to both sides, which is the square of half of the coefficient of the middle term $-6x$. The current equation is $x^2 - 6x + 9 = 7$. The left side can be factored into a square of a binomial, resulting in $(x - 3)^2 = 7$. To solve for x, the square root of both sides should be taken, resulting in $(x - 3) = \pm\sqrt{7}$, and $x = 3 \pm \sqrt{7}$.

Other ways of solving quadratic equations include graphing, factoring, and using the quadratic formula. The equation $y = x^2 - 4x + 3$ can be graphed on the coordinate plane, and the solutions can be observed where it crosses the x-axis. The graph will be a parabola that opens up with two solutions at 1 and 3.

If quadratic equations take the form $ax^2 - b = 0$, then the equation can be solved by adding *b* to both sides and dividing by *a* to get $x^2 = \frac{b}{a}$ or $x = \pm\sqrt{\frac{b}{a}}$. Note that this is actually two separate solutions, unless *b* happens to be 0.

If a quadratic equation has no constant—so that it takes the form $ax^2 + bx = 0$—then the *x* can be factored out to get $x(ax + b) = 0$. Then, the solutions are $x = 0$, together with the solutions to $ax + b = 0$. Both factors $x \text{ } and \text{ } (ax + b)$ can be set equal to zero to solve for x because one of those values must be zero for their product to equal zero. For an equation $ab = 0$ to be true, either $a = 0$, or $b = 0$.

A given quadratic equation $x^2 + bx + c$ can be factored into $(x + A)(x + B)$, where $A + B = b$, and $AB = c$. Finding the values of A and B can take time, but such a pair of numbers can be found by guessing and checking. Looking at the positive and negative factors for *c* offers a good starting point.

For example, in $x^2 - 5x + 6$, the factors of 6 are 1, 2, and 3. Now,$(-2)(-3) = 6$, and $-2 - 3 = -5$. In general, however, this may not work, in which case another approach may need to be used.

A quadratic equation of the form $x^2 + 2xb + b^2 = 0$ can be factored into $(x + b)^2 = 0$. Similarly, $x^2 - 2xy + y^2 = 0$ factors into $(x - y)^2 = 0$.

The first method of completing the square can be used in finding the second method, the quadratic formula. It can be used to solve any quadratic equation. This formula may be the longest method for solving quadratic equations and is commonly used as a last resort after other methods are ruled out.

It can be helpful in memorizing the formula to see where it comes from, so here are the steps involved:

The most general form for a quadratic equation is $ax^2 + bx + c = 0$.

First, dividing both sides by *a* leaves us with $x^2 + \frac{b}{a}x + \frac{c}{a} = 0$.

To complete the square on the left-hand side, c/a can be subtracted on both sides to get $x^2 + \frac{b}{a}x = -\frac{c}{a}$.

$(\frac{b}{2a})^2$ is then added to both sides.

This gives $x^2 + \frac{b}{a}x + (\frac{b}{2a})^2 = (\frac{b}{2a})^2 - \frac{c}{a}$.

The left can now be factored and the right-hand side simplified to give $(x + \frac{b}{2a})^2 = \frac{b^2-4ac}{4a}$.

Taking the square roots gives $x + \frac{b}{2a} = \pm\frac{\sqrt{b^2-4ac}}{2a}$.

Solving for x yields the quadratic formula: $x = \frac{-b\pm\sqrt{b^2-4ac}}{2a}$.

It isn't necessary to remember how to get this formula, but memorizing the formula itself is the goal.

If an equation involves taking a root, then the first step is to move the root to one side of the equation and everything else to the other side. That way, both sides can be raised to the index of the radical in order to remove it, and solving the equation can continue.

Exponential Functions

The general form for an **exponential function** is $y = ab^x$ where *b* must be a positive number and cannot equal 1. When the value of *b* is greater than 1, the function models exponential **growth** (as *x* increases, *y* increases). When the value of *b* is less than 1, the function models exponential **decay** (as *x* increases, y decreases). If a is positive, the graph consists of points above the x-axis; if a is negative, the graph consists of points below the x-axis. An asymptote is a line that a graph approaches.

Exponential Curve

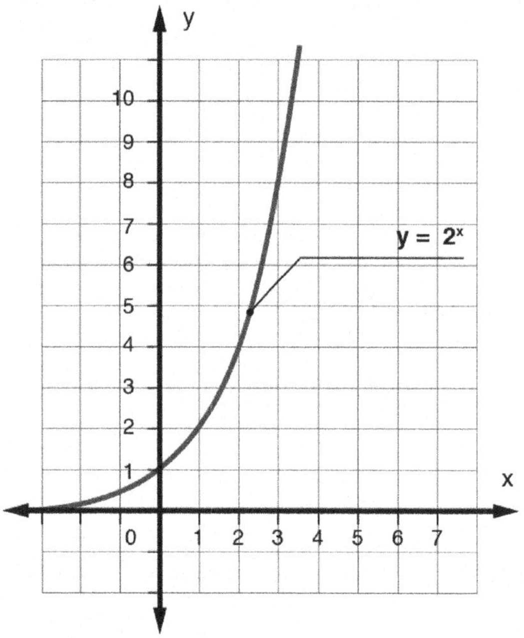

To solve an equation involving exponential expressions, the goal is to isolate the exponential expression. Once this process is completed, the logarithm—with the base equaling the base of the exponent of both sides—needs to be taken to get an expression for the variable. If the base is e, the natural log of both sides needs to be taken.

To solve an equation with logarithms, the given equation needs to be written in exponential form, using the fact that $\log_b y = x$ means $b^x = y$, and then solved for the given variable. Lastly, properties of logarithms can be used to simplify more than one logarithmic expression into one.

When working with logarithmic functions, it is important to remember the following properties. Each one can be derived from the definition of the logarithm as the inverse to an exponential function:

$$\log_b 1 = 0$$

$$\log_b b = 1$$

$$\log_b b^p = p$$

$$\log_b MN = \log_b M + \log_b N$$

$$\log_b \frac{M}{N} = \log_b M - \log_b N$$

$$\log_b M^p = p \log_b M$$

When solving equations involving exponentials and logarithms, the following fact should be used:

If f is a one-to-one function, $a = b$ is equivalent to $f(a) = f(b)$.

Using this, together with the fact that logarithms and exponentials are inverses, allows manipulations of the equations to isolate the variable.

Example:

Solve $4 = \ln(x - 4)$.

Using the definition of a logarithm, the equation can be changed to $e^4 = e^{\ln(x-4)}$.

The functions on the right side cancel with a result of $e^4 = x - 4$.

This then gives $x = 4 + e^4$.

Systems of Equations and Inequalities

Representing and Solving Systems of Linear Equations and Inequalities with Two Variables Algebraically and Graphically

A **system of equations** is a group of equations that have the same variables or unknowns. These equations can be linear, but they are not always so. Finding a solution to a system of equations means finding the values of the variables that satisfy each equation. For a linear system of two equations and two variables, there could be a single solution, no solution, or infinitely many solutions.

A single solution occurs when there is one value for x and y that satisfies the system. This would be shown on the graph where the lines cross at exactly one point. When there is no solution, the lines are parallel and do not ever cross. With infinitely many solutions, the equations may look different, but they are the same line. One equation will be a multiple of the other, and on the graph, they lie on top of each other.

The process of elimination can be used to solve a system of equations. For example, the following equations make up a system:

$$x + 3y = 10 \text{ and } 2x - 5y = 9$$

Immediately adding these equations does not eliminate a variable, but it is possible to change the first equation by multiplying the whole equation by -2. This changes the first equation to

$$-2x - 6y = -20$$

The equations can be then added to obtain $-11y = -11$. Solving for y yields $y = 1$. To find the rest of the solution, 1 can be substituted in for y in either original equation to find the value of $x = 7$. The solution to the system is (7, 1) because it makes both equations true, and it is the point in which the lines intersect. If the system is **dependent**—having infinitely many solutions—then both variables will cancel out when the elimination method is used, resulting in an equation that is true for many values of *x* and *y*. Since the system is dependent, both equations can be simplified to the same equation or line.

A system can also be solved using **substitution.** This involves solving one equation for a variable and then plugging that solved equation into the other equation in the system. For example, $x - y = -2$ and

$3x + 2y = 9$ can be solved using substitution. The first equation can be solved for x, where $x = -2 + y$. Then it can be plugged into the other equation:

$$3(-2 + y) + 2y = 9$$

Solving for y yields:

$$-6 + 3y + 2y = 9$$

That shows that $y = 3$. If $y = 3$, then $x = 1$.

This solution can be checked by plugging in these values for the variables in each equation to see if it makes a true statement.

Finally, a solution to a system of equations can be found graphically. The solution to a linear system is the point or points where the lines cross. The values of x and y represent the coordinates (x, y) where the lines intersect. Using the same system of equations as above, they can be solved for y to put them in slope-intercept form, $y = mx + b$. These equations become $y = x + 2$ and $y = -\frac{3}{2}x + 4.5$. The slope is the coefficient of x, and the y-intercept is the constant value.

This system with the solution is shown below:

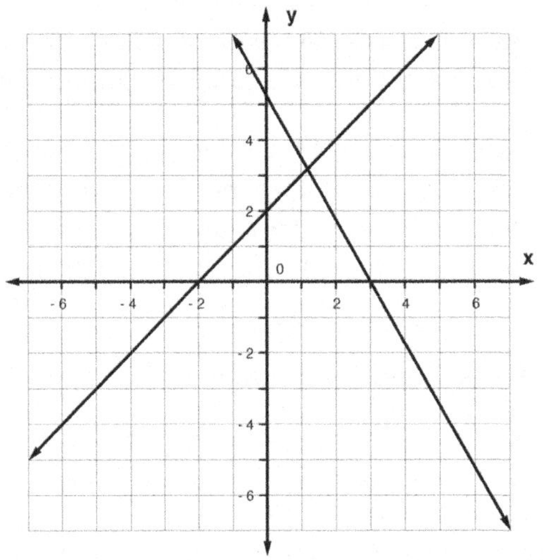

A solution for a system of equations is an ordered pair that makes both equations true. Graphing both lines on a coordinate plane can help find the solution to the system. If the lines intersect, the point of intersection is the solution to the system. Every point on a line represents an ordered pair that makes its equation true. The ordered pair represented by this point of intersection lies on both lines and therefore makes both equations true. This ordered pair should be checked by substituting its values into both of the original equations of the system. Note that given a system of equations and an ordered pair, the ordered pair can be determined to be a solution or not by checking it in both equations.

If, when graphed, the lines representing the equations of a system do not intersect, then the two lines are parallel to each other or they are the same exact line. Parallel lines extend in the same direction without ever meeting. A system consisting of parallel lines has no solution. If the equations for a system represent the same exact line, then every point on the line is a solution to the system. In this case, there would be an infinite number of solutions. A system consisting of intersecting lines is referred to as independent; a system consisting of parallel lines is referred to as inconsistent; and a system consisting of coinciding lines is referred to as dependent.

Parallel Lines	**Intersecting Lines**	**Coincident Lines**
Inconsistent	Independent	Dependent

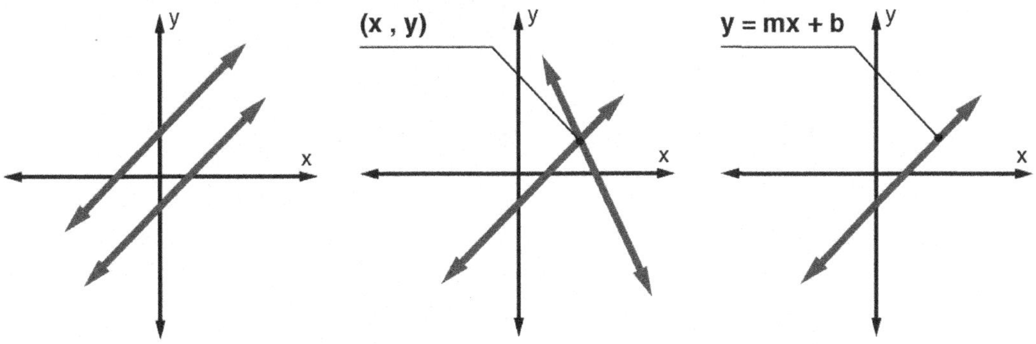

Systems of linear inequalities are like systems of equations, but the solutions are different. Since inequalities have infinitely many solutions, their systems also have infinitely many solutions. Finding the solutions of inequalities involves graphs. A system of two equations and two inequalities is linear; thus, the lines can be graphed using slope-intercept form. If the inequality has an equals sign, the line is solid. If the inequality only has a greater than or less than symbol, the line on the graph is dotted. Dashed lines indicate that points lying on the line are not included in the solution. After the lines are graphed, a region is shaded on one side of the line. This side is found by determining if a point—known as a **test point**—lying on one side of the line produces a true inequality. If it does, that side of the graph is shaded. If the point produces a false inequality, the line is shaded on the opposite side from the point. The graph of a system of inequalities involves shading the intersection of the two shaded regions.

A system of linear inequalities consists of two linear inequalities making comparisons between two variables. Students may be given a scenario and asked to express it as a system of inequalities:

> A consumer study calls for at least 60 adult participants. It cannot use more than 25 men. Express these constraints as a system of inequalities.

This can be modeled by the system: $x + y \geq 60; x \leq 25$, where *x* represents the number of men and *y* represents the number of women. A solution to the system is an ordered pair that makes both inequalities true when substituting the values for *x* and *y*.

The solution set for a system of inequalities is the region of a graph consisting of ordered pairs that make both inequalities true. To graph the solution set, each linear inequality should first be graphed with appropriate shading. The region of the graph should be identified where the shading for the two inequalities overlaps. This region contains the solution set for the system.

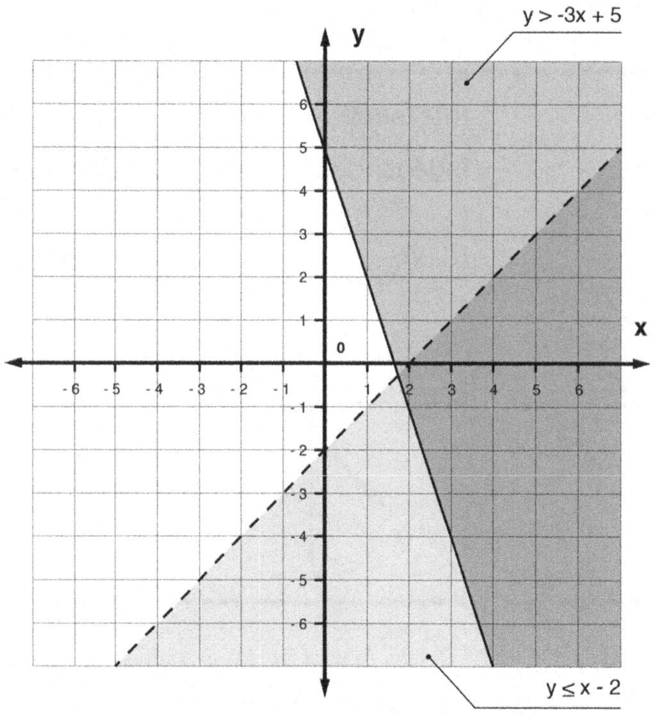

An ordered pair from the region of solutions can be selected to test in the system of inequalities.

Just as with manipulating linear inequalities in one variable, if dividing by a negative number in working with a linear inequality in two variables, the relationship reverses and the inequality sign should be flipped.

Recognizing and Representing Simple Sequences or Patterns

Evaluating, Extending, or Algebraically Representing Rules that Involve Number Patterns

Patterns are an important part of mathematics. When mathematical calculations are completed repeatedly, patterns can be recognized. Recognizing patterns is an integral part of mathematics because it helps you understand relationships between different ideas. For example, a sequence of numbers can be given, and being able to recognize the relationship between the given numbers can help in completing the sequence.

For instance, given the sequence of numbers $7, 14, 21, 28, 35, \ldots$, the next number in the sequence would be 42. This is because the sequence lists all multiples of 7, starting at 7. Sequences can also be built from addition, subtraction, and division. Being able to recognize the relationship between the values that are given is the key to finding out the next number in the sequence.

Patterns within a sequence can come in 2 distinct forms. The items either repeat in a constant order, or the items change from one step to another in some consistent way. The **core** is the smallest unit, or number of items, that repeats in a repeating pattern. For example, the pattern oo▲oo▲o... has a core that is oo▲. Knowing only the core, the pattern can be extended. Knowing the number of steps in the core allows the identification of an item in each step without drawing/writing the entire pattern out. For example, suppose you must find the tenth item in the previous pattern. Because the core consists of three items (○○▲), the core repeats in multiples of 3. In other words, steps 3, 6, 9, 12, etc. will be ▲ completing the core with the core starting over on the next step. For the above example, the 9th step will be ▲ and the 10th will be ○.

The most common patterns where each item changes from one step to the next are arithmetic and geometric sequences. In an **arithmetic sequence**, the items increase or decrease by a constant difference. In other words, the same thing is added or subtracted to each item or step to produce the next. To determine if a sequence is arithmetic, see what must be added or subtracted to step one to produce step two. Then, check if the same thing is added/subtracted to step two to produce step three. The same thing must be added/subtracted to step three to produce step four, and so on. Consider the pattern 13, 10, 7, 4, … . To get from step one (13) to step two (10) by adding or subtracting requires subtracting by 3. The next step is checking if subtracting 3 from step two (10) will produce step three (7), and subtracting 3 from step three (7) will produce step four (4). In this case, the pattern holds true. Therefore, this is an arithmetic sequence in which each step is produced by subtracting 3 from the previous step. To extend the sequence, 3 is subtracted from the last step to produce the next. The next three numbers in the sequence are 1, -2, -5.

A **geometric sequence** is one in which each step is produced by multiplying or dividing the previous step by the same number. To see if a sequence is geometric, decide what step one must be multiplied or divided by to produce step two. Then check if multiplying or dividing step two by the same number produces step three, and so on. Consider the pattern 2, 8, 32, 128, … . To get from step one (2) to step two (8) requires multiplication by 4. The next step determines if multiplying step two (8) by 4 produces step three (32), and multiplying step three (32) by 4 produces step four (128). In this case, the pattern holds true. Therefore, this is a geometric sequence in which each step is found by multiplying the previous step by 4. To extend the sequence, the last step is multiplied by 4 and repeated. The next three numbers in the sequence are 512; 2,048; 8,192.

Patterns Involving Shapes or Figures

Arithmetic and geometric sequences can also be represented by shapes. For example, an arithmetic sequence could consist of shapes with three sides, four sides, and five sides. A geometric sequence could consist of eight blocks, four blocks, and two blocks (each step is produced by dividing the number of blocks in the previous step by 2).

Using Patterns to Make Conjectures, Predictions, or Generalizations

Patterns can be used to make conjectures, predictions, or generalizations. One of the most common question types test takers will encounter is to identify what number or shape will come next, which involves making a prediction based on understanding the presented pattern.

When given two number patterns, the corresponding terms should be examined to determine if a relationship exists between them. Corresponding terms between patterns are the pairs of numbers which appear in the same step of the two sequences. Consider the following patterns 1, 2, 3, 4,… and 3, 6, 9, 12, … . The corresponding terms are: 1 and 3; 2 and 6; 3 and 9; and 4 and 12. To identify the relationship, each pair of corresponding terms is examined. You can also examine the possibilities of performing an operation (+, −, ×, ÷) to each sequence.

In this case:

$1 + 2 = 3$ or $1 \times 3 = 3$

$2 + 4 = 6$ or $2 \times 3 = 6$

$3 + 6 = 9$ or $3 \times 3 = 9$

$4 + 8 = 12$ or $4 \times 3 = 12$

The pattern is that the number from the first sequence multiplied by 3 equals the number in the second sequence. By assigning each sequence a label (input and output) or variable (*x* and *y*), the relationship can be written as an equation. The first sequence represents the inputs, or *x*, and the second sequence represents the outputs, or *y*. So, the relationship can be expressed as: $y = 3x$.

Consider the following sets of numbers:

a	2	4	6	8
b	6	8	10	12

To write a rule for the relationship between the values for *a* and the values for *b*, the corresponding terms (2 and 6; 4 and 8; 6 and 10; 8 and 12) are examined. The possibilities for producing *b* from *a* are:

$2 + 4 = 6$ or $2 \times 3 = 6$

$4 + 4 = 8$ or $4 \times 2 = 8$

$6 + 4 = 10$

$8 + 4 = 12$ or $8 \times 1.5 = 12$

The pattern is that adding 4 to the value of *a* produces the value of *b*. The relationship can be written as the equation $a + 4 = b$.

Functions and Their Graphs

Identifying, Defining, and Evaluating Functions

Function Notation

A **function** is defined as a relationship between inputs and outputs where there is only one output value for a given input.

A function can be viewed as an object that has *x* as its input and outputs a unique *y*-value. It is sometimes convenient to express this using **function notation**, where the function itself is given a name, often *f*. To emphasize that *f* takes *x* as its input, the function is written as $f(x)$. In the above example, the equation could be rewritten as $f(x) = x^2 + 1$. To write the value that a function yields for some specific value of *x*, that value is put in place of *x* in the function notation. For example, $f(3)$ means the value that the function outputs when the input value is 3. If $f(x) = x^2 + 1$, then $f(3) = 3^2 + 1 = 10$.

Another example of a function would be $f(x) = 4x + 4$, read "*f* of *x* is equal to four times *x* plus four." In this example, the input would be *x* and the output would be f(*x*). Ordered pairs would be represented as (*x*, f(*x*)). To find the output for an input value of 3, 3 would be substituted for *x* into the function as follows: $f(3) = 4(3) + 4$, resulting in $f(3) = 16$. Therefore, the ordered pair $(3, f(3)) = (3, 16)$. Note f(*x*) is a function of *x* denoted by *f*. Functions of *x* could be named *g*(*x*), read "*g* of *x*"; *p*(*x*), read "*p* of *x*"; etc.

As an example, the following function is in function notation: $f(x) = 3x - 4$. The $f(x)$ represents the output value for an input of x. If $x = 2$, the equation becomes:

$$f(2) = 3(2) - 4 = 6 - 4 = 2$$

The input of 2 yields an output of 2, forming the ordered pair $(2, 2)$. The following set of ordered pairs corresponds to the given function: $(2, 2), (0, -4), (-2, -10)$. The set of all possible inputs of a function is its **domain**, and all possible outputs is called the **range**. By definition, each member of the domain is paired with only one member of the range.

Functions can also be defined recursively. In this form, they are not defined explicitly in terms of variables. Instead, they are defined using previously-evaluated function outputs, starting with either $f(0)$ or $f(1)$. An example of a recursively-defined function is:

$$f(1) = 2, f(n) = 2f(n-1) + 2n, n > 1$$

The domain of this function is the set of all integers.

A function can also be viewed as a table of pairs (*x*, *y*), which lists the value for *y* for each possible value of *x*.

Deciding Whether a Given Set of Conditions Represents a Function

First, it's important to understand the definition of a **relation**. Given two variables, *x* and *y*, which stand for unknown numbers, a *relation* between *x* and *y* is an object that splits all of the pairs (*x*, *y*) into those for which the relation is true and those for which it is false. For example, consider the relation of $x^2 = y^2$. This relationship is true for the pair (1, 1) and for the pair (-2, 2), but false for (2, 3). Another example of a relation is $x \leq y$. This is true whenever *x* is less than or equal to *y*.

A **function** is a special kind of relation where, for each value of *x*, there is only a single value of *y* that satisfies the relation. So, $x^2 = y^2$ is *not* a function because in this case, if *x* is 1, *y* can be either 1 or -1: the pair (1, 1) and (1, -1) both satisfy the relation. More generally, for this relation, any pair of the form $(a, \pm a)$ will satisfy it. On the other hand, consider the following relation: $y = x^2 + 1$. This is a function because for each value of *x*, there is a unique value of *y* that satisfies the relation. Notice, however, there are multiple values of *x* that give us the same value of *y*. This is perfectly acceptable for a function. Therefore, *y* is a function of *x*.

To determine if a relation is a function, check to see if every *x* value has a unique corresponding *y* value.

Evaluating Functions for Given Values (algebraically, graphically, tabular)

To evaluate functions, plug in the given value everywhere the variable appears in the expression for the function. For example, find $f(-2)$ where $f(x) = 2x^2 - \frac{4}{x}$. To complete the problem, plug in -2 in the following way:

$$f(-2) = 2(-2)^2 - \frac{4}{-2} = 2 \times 4 + 2 = 8 + 2 = 10$$

The Domain and the Range of a Function

Determining the Domain and Range of a Given Table of Values

The domain and range of a function can be found by observing a table. The table below shows the input values $x = -2$ to $x = 2$ for the function $f(x) = x^2 - 3$. The range, or output, for these inputs results in a minimum of -3. On each side of $x = 0$, the numbers increase, showing that the range is all real numbers greater than or equal to -3.

x (domain/input)	y (range/output)
-2	1
-1	-2
0	-3
-1	-2
2	1

Determining the Domain and Range from a Given Graph of a Function

The domain and range of a function can also be found visually by its plot on the coordinate plane. In the function $f(x) = x^2 - 3$, for example, the domain is all real numbers because the parabola can stretch infinitely far left and right with no restrictions. This means that any input value from the real number system will yield an output in the real number system. For the range, the inequality $y \geq -3$ would be used to describe the possible output values because the parabola has a minimum at $y = -3$. This means there will not be any real output values less than -3 because -3 is the lowest value the function reaches on the y-axis.

Determining the Domain and Range of a Given Function

The set of all possible values for x in $f(x)$ is called the **domain** of the function, and the set of all possible outputs is called the **range** of the function. Note that usually the domain is assumed to be all real numbers, except those for which the expression for $f(x)$ is not defined, unless the problem specifies otherwise. An example of how a function might not be defined is in the case of $f(x) = \frac{1}{x+1}$, which is not defined when $x = -1$ (which would require dividing by zero). Therefore, in this case the domain would be all real numbers except $x = -1$.

Interpreting Domain and Range in Real-World Settings

A function can be built from the information given in a situation. For example, the relationship between the money paid for a gym membership and the number of months that someone has been a member can be described through a function. If the one-time membership fee is $40 and the monthly fee is $30, then the function can be written $f(x) = 30x + 40$. The x-value represents the number of months the person has been part of the gym, while the output is the total money paid for the membership. The table below shows this relationship. It is a representation of the function because the initial cost is $40 and the cost increases each month by $30.

x (months)	f(x) (money paid to gym)
0	40
1	70
2	100
3	130

In this situation, the domain of the function is real numbers greater than or equal to zero because it represents the number of months that a membership is held. We aren't told if the gym prorates memberships for partial months (if you join 10 days into a month, for example). If not, the domain would only be whole numbers plus zero, since there's a meaningful data point of $40, as a fee for joining. The range is real numbers greater than or equal to 40, because the range represents the total cost of the gym membership. Because there is a one-time fee of $40, the cost of carrying a membership will never be less than $40, so this is the minimum value.

When working through any word problem, the domain and range of the function should be considered in terms of the real-world context that the function models. For example, considering the above function for the cost of a gym membership, it would be nonsensical to include negative numbers in either the domain or range because there can't be negative months that someone holds a membership and similarly, the gym isn't going to pay a person for months prior to becoming a member. Therefore, while the function to model the situation (defined as $f(x) = 30x + 40$) theoretically could result in a true mathematical statement if negative values of are inputted, this would not make sense in the real-world context for which the function applies. Therefore, defining the domain as whole numbers and the range as all real numbers greater than or equal to 40 is important.

The Basic Characteristics of Linear Functions

The Slope of a Linear Function

Rate of change for any line calculates the steepness of the line over a given interval. Rate of change is also known as the **slope** or rise/run. The slope of a linear function is given by the change in *y* divided by the change in *x*. So the formula looks like this:

$$slope = \frac{y_2 - y_1}{x_2 - x_1}$$

In the graph below, two points are plotted. The first has the coordinates of (0, 1), and the second point is (2, 3). Remember that the x coordinate is always placed first in coordinate pairs. Work from left to right when identifying coordinates. Thus, the point on the left is point 1 (0, 1), and the point on the right is point 2 (2, 3).

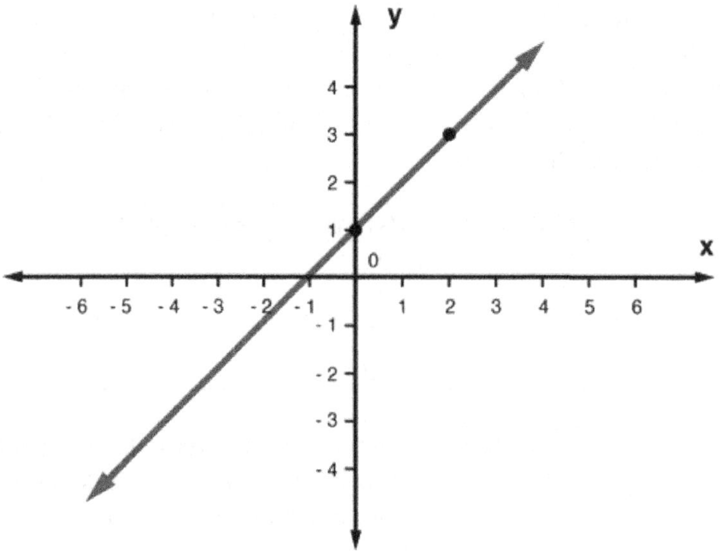

Now we need to just plug those numbers into the equation:

$$slope = \frac{3-1}{2-0}$$

$$slope = \frac{2}{2}$$

$$slope = 1$$

This means that for every increase of 1 for x, y also increased by 1. You can see this in the line. When x equaled 0, y equaled 1, and when x was increased to 1, y equaled 2.

Slope can be thought of as determining the rise over run:

$$slope = \frac{rise}{run}$$

The rise being the change vertically on the y axis and the run being the change horizontally on the x axis.

Interpreting Slope as a Constant Rate of Change

The first step to writing a linear model is to identify what the variables represent. A variable represents an unknown quantity, and in the case of a linear equation, a specific relationship exists between the two variables (usually *x* and *y*). Within a given scenario, the variables are the two quantities that are changing. The variable *x* is considered the independent variable and represents the inputs of a function. The variable *y* is considered the dependent variable and represents the outputs of a function. For example, if a scenario describes distance traveled and time traveled, distance would be represented by *y* and time represented by *x*. The distance traveled depends on the time spent traveling (time is independent). If a scenario describes the cost of a cab ride and the distance traveled, the cost would be represented by *y* and the distance represented by *x*. The cost of a cab ride depends on the distance traveled.

As mentioned, the slope represents the rate of change. In linear models, because the slope is constant (a straight line), the rate of change is also constant. In the case of the cab ride, as the distance of the ride increases, so too does the cost in a directly proportional rate.

Determining the *x*- and *y*-Intercepts of a Given Linear Function

The x- and y-intercepts of a given linear function can be determined based on the form that the linear function is presented in. Of course, if the function is graphed, the x-intercept is the point where the graph of the line crosses the x-axis, while the y-intercept is the point where the graph of the line crosses the y-axis.

It is easiest to determine the intercepts of a linear function when it is written in intercept form ($\frac{x}{x_1} + \frac{y}{y_1} = 1$). Here, x_1 is the *x*-intercept and y_1 is the *y*-intercept. When a linear function is written in standard form ($Ax + By = C$), the *y*-intercept is given by $\frac{C}{B}$. The y-intercept of a linear equations written in slope-intercept form ($y = mx + b$) is *b*.

Interpreting the *x*- and *y*-Intercepts of a Given Linear Function

The slope of the graph of a line represents the rate of change between the variables of an equation. In the context of a real-world scenario, the slope will tell the way in which the unknown quantities (variables) change with respect to each other. A scenario involving distance and time might state that someone is traveling at a rate of 45 miles per hour. The slope of the linear model would be 45. A scenario involving the cost of a cab ride and distance traveled might state that the person is charged $3 for each mile. The slope of the linear model would be 3.

The *y*-intercept of a linear function is the value of *y* when $x = 0$ (the point where the line intercepts the *y*-axis on the graph of the equation). It is sometimes helpful to think of this as a "starting point" for a linear function. Suppose for the scenario about the cab ride that the person is told that the cab company charges a flat fee of $5 plus $3 for each mile. Before traveling any distance ($x = 0$), the cost is $5. The *y*-intercept for the linear model would be 5.

A linear equation with two variables can be written given a point (ordered pair) and the slope or given two points on a line. An ordered pair gives a set of corresponding values for the two variables (*x* and *y*). As an example, for a scenario involving distance and time, it is given that the person traveled 112.5 miles in 2 ½ hours. Knowing that *x* represents time and *y* represents distance, this information can be written as the ordered pair (2.5, 112.5).

The Relationships Among Functions, Tables, and Graphs

Determining and Interpreting the *x*- and *y*-Intercepts of Any Given Function

As mentioned, linear relationships describe the way two quantities change with respect to each other. The relationship is defined as linear because a line is produced if all the sets of corresponding values are graphed on a coordinate grid. When expressing the linear relationship as an equation, the equation is often written in the form $y = mx + b$ (**slope-intercept form**) where *m* and *b* are numerical values and *x* and *y* are variables (for example, $y = 5x + 10$). The slope is the coefficient of x, and the y-intercept is the constant value. The slope of the line containing the same two points is $m = \frac{y_2 - y_1}{x_2 - x_1}$ and is also equal to rise/run. Given a linear equation and the value of either variable (*x* or *y*), the value of the other variable can be determined.

With polynomial functions such as quadratics, the x-intercepts represent zeros of the function. Finding the **zeros of polynomial functions** is the same process as finding the solutions of polynomial equations. These are the points at which the graph of the function crosses the x-axis. In the following quadratic equation, factoring the binomial leads to finding the zeros of the function: $x^2 - 5x + 6 = y$. This equations factors into $(x - 3)(x - 2) = y$, where 2 and 3 are found to be the zeros of the function when y is set equal to zero. The zeros of any function are the x-values where the graph of the function on the coordinate plane crosses the x-axis, which is the same as an x-intercept.

Selecting an Equation that Best Represents a Graph

Three common functions used to model different relationships between quantities are linear, quadratic, and exponential functions. **Linear functions** are the simplest of the three, and the independent variable x has an exponent of 1. Written in the most common form, $y = mx + b$, the coefficient of x indicates how fast the function grows at a constant rate, and the b-value denotes the starting point. A **quadratic function** has an exponent of 2 on the independent variable x. Standard form for this type of function is $y = ax^2 + bx + c$, and the graph is a parabola. These type functions grow at a changing rate. **An exponential function** has an independent variable in the exponent $y = ab^x$. The graph of these types of functions is described as **growth** or **decay**, based on whether the **base**, b, is greater than or less than 1. These functions are different from quadratic functions because the base stays constant. A common base is base *e*.

The following three functions model a linear, quadratic, and exponential function respectively: $y = 2x$, $y = x^2$, and $y = 2^x$. Their graphs are shown below. The first graph, modeling the linear function, shows that the growth is constant over each interval. With a horizontal change of 1, the vertical change is 2. It models a constant positive growth. The second graph shows the quadratic function, which is a curve that is symmetric across the y-axis. The growth is not constant, but the change is mirrored over the axis. The last graph models the exponential function, where the horizontal change of 1 yields a vertical change that increases more and more. The exponential graph gets very close to the x-axis, but never touches it, meaning there is an asymptote there. The y-value can never be zero because the base of 2 can never be raised to an input value that yields an output of zero.

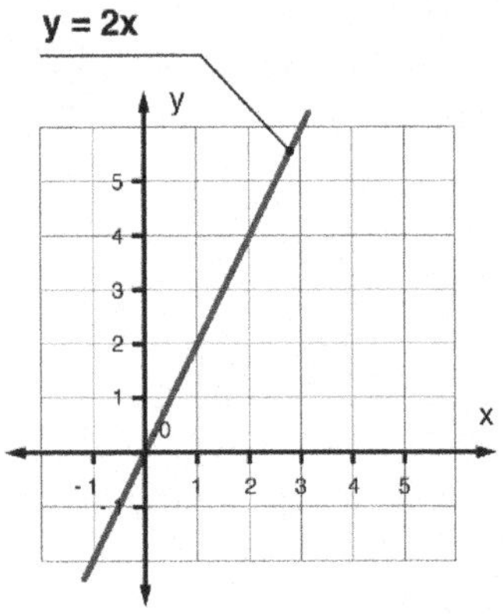

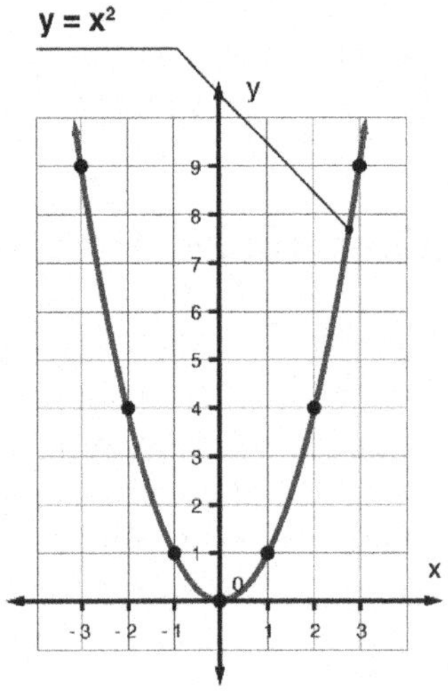

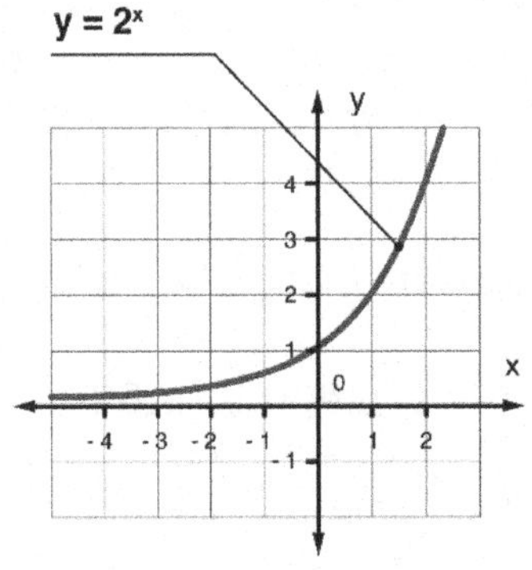

Determining the Graphical Properties and Sketch a Graph Given an Equation

Graphing a Linear Function

The process for graphing a line depends on the form in which its equation is written: slope-intercept form or standard form.

When an equation is written in slope-intercept form, $y = mx + b$, m represents the slope of the line and b represents the y-intercept. The y-intercept is the value of y when $x = 0$ and the point at which the graph of the line crosses the y-axis. The slope is the rate of change between the variables, expressed as a fraction. The fraction expresses the change in y compared to the change in x. If the slope is an integer, it should be written as a fraction with a denominator of 1. For example, 5 would be written as 5/1.

To graph a line given an equation in slope-intercept form, the y-intercept should first be plotted. For example, to graph $y = -\frac{2}{3}x + 7$, the y-intercept of 7 would be plotted on the y-axis (vertical axis) at the point (0, 7). Next, the slope would be used to determine a second point for the line. Note that all that is necessary to graph a line is two points on that line. The slope will indicate how to get from one point on the line to another. The slope expresses vertical change (y) compared to horizontal change (x) and therefore is sometimes referred to as $\frac{rise}{run}$. The numerator indicates the change in the y value (move up for positive integers and move down for negative integers), and the denominator indicates the change in the x value. For the previous example, using the slope of $-\frac{2}{3}$, from the first point at the y-intercept, the second point should be found by counting down 2 and to the right 3. This point would be located at (3, 5).

When an equation is written in standard form, $Ax + By = C$, it is easy to identify the x- and y-intercepts for the graph of the line. Just as the y-intercept is the point at which the line intercepts the y-axis, the x-intercept is the point at which the line intercepts the x-axis. At the y-intercept, $x = 0$; and at the x-intercept, $y = 0$. Given an equation in standard form, $x = 0$ should be used to find the y-intercept. Likewise, $y = 0$ should be used to find the x-intercept. For example, to graph $3x + 2y = 6$, 0 for y results in $3x + 2(0) = 6$. Solving for y yields $x = 2$; therefore, an ordered pair for the line is (2, 0). Substituting 0 for x results in $3(0) + 2y = 6$. Solving for y yields $y = 3$; therefore, an ordered pair for the line is (0, 3). The two ordered pairs (the x- and y-intercepts) can be plotted, and a straight line through them can be constructed.

Take a look at this graphic:

T - chart

x	y
0	3
2	0

Intercepts

x - intercept : (2,0)

y - intercept : (0,3)

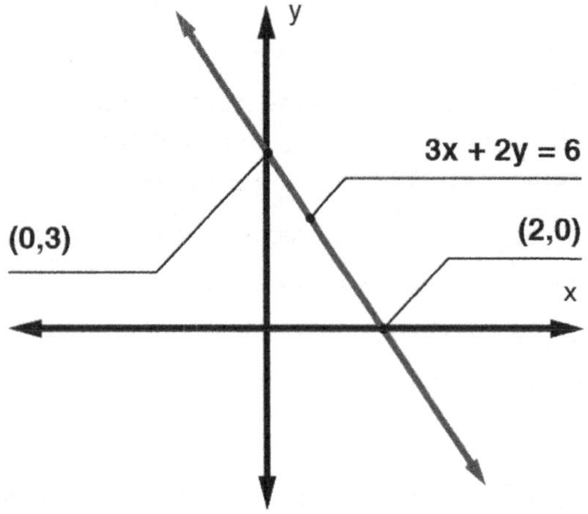

Graphing a Quadratic Function

The standard form of a quadratic function is $y = ax^2 + bx + c$. The graph of a quadratic function is a u-shaped (or upside down u) curve, called a parabola, which is symmetric about a vertical line (axis of symmetry). To graph a parabola, its vertex (high or low point for the curve) and at least two points on each side of the axis of symmetry need to be determined.

Given a quadratic function in standard form, the axis of symmetry for its graph is the line $x = -\frac{b}{2a}$. The vertex for the parabola has an *x*-coordinate of $-\frac{b}{2a}$. To find the *y*-coordinate for the vertex, the calculated *x*-coordinate needs to be substituted. To complete the graph, two different *x*-values need to be selected and substituted into the quadratic function to obtain the corresponding *y*-values. This will give two points on the parabola. These two points and the axis of symmetry are used to determine the two points corresponding to these. The corresponding points are the same distance from the axis of symmetry (on the other side) and contain the same *y*-coordinate. Plotting the vertex and four other points on the parabola allows for constructing the curve.

Take a look at this graphic:

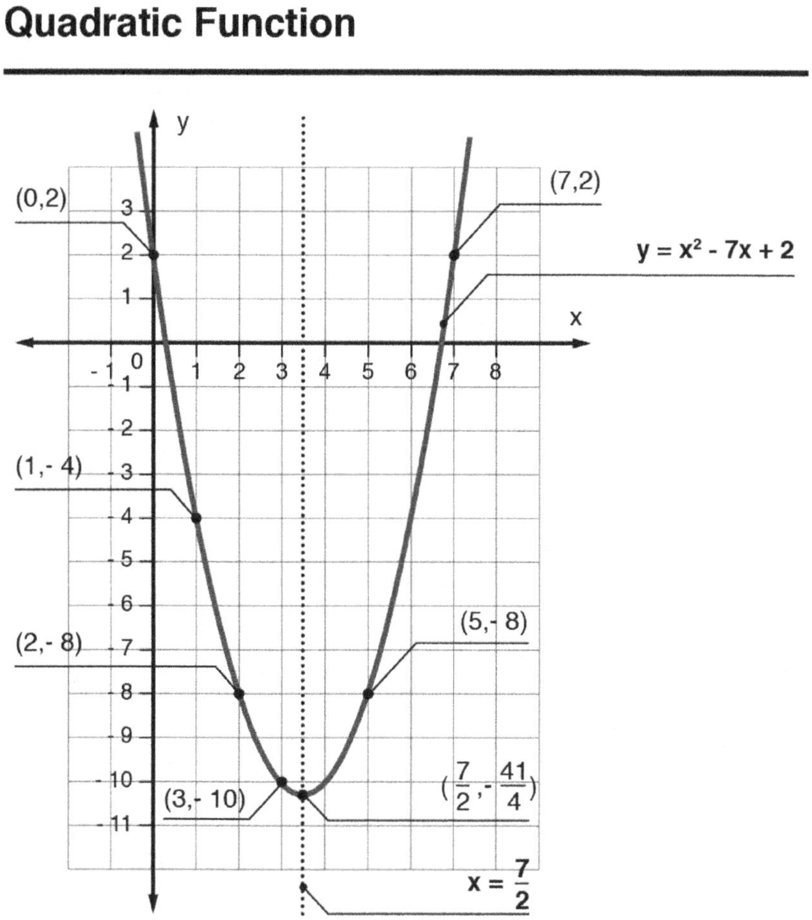

Graphing an Exponential Function

Exponential functions have a general form of $y = a \text{ x } b^x$. The graph of an exponential function is a curve that slopes upward or downward from left to right. The graph approaches a line, called an asymptote, as *x* or *y* increases or decreases. To graph the curve for an exponential function, *x*-values are selected and then substituted into the function to obtain the corresponding *y*-values. A general rule of thumb is to select three negative values, zero, and three positive values. Plotting the seven points on the graph for an exponential function should allow for constructing a smooth curve through them.

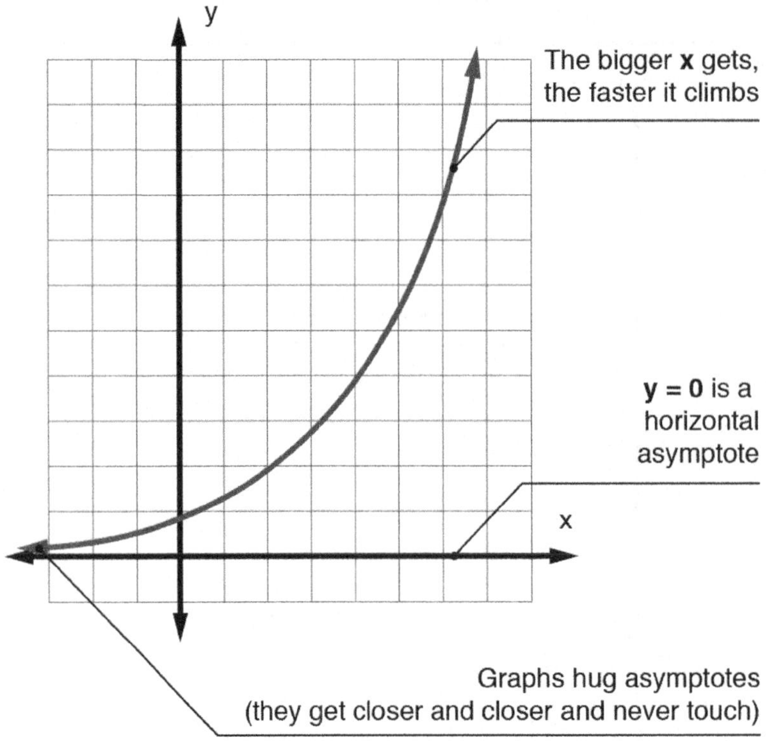

Analyzing and Representing Functions that Model Given Information

Developing a Model of a Given Set of Conditions

The three tables below show specific values for three types of functions. The third column in each table shows the change in the y-values for each interval. The first table shows a constant change of 2 for each equal interval, which matches the slope in the equation $y = 2x$. The second table shows an increasing change, but it also has a pattern. The increase is changing by 2 more each time, so the change is quadratic. The third table shows the change as factors of the base, 2. It shows a continuing pattern of factors of the base.

$y = 2x$		
x	y	Δy
1	2	
2	4	2
3	6	2
4	8	2
5	10	2

$y = x^2$		
x	y	Δy
1	1	
2	4	3
3	9	5
4	16	7
5	25	9

$y = 2^x$		
x	y	Δy
1	2	
2	4	2
3	8	4
4	16	8
5	32	16

Given a table of values, the type of function can be determined by observing the change in y over equal intervals. For example, the tables below model two functions. The changes in interval for the x-values is 1 for both tables. For the first table, the y-values increase by 5 for each interval. Since the change is constant, the situation can be described as a linear function. The equation would be $y = 5x + 3$. For the second table, the change for y is 5, 20, 100, and 500, respectively. The increases are multiples of 5, meaning the situation can be modeled by an exponential function. The equation $y = 5^x + 3$ models this situation.

x	y
0	3
1	8
2	13
3	18
4	23

x	y
0	3
1	8
2	28
3	128
4	628

Quadratic equations can be used to model real-world area problems. For example, a farmer may have a rectangular field that he needs to sow with seed. The field has length $x + 8$ and width $2x$. The formula for area should be used: $A = lw$. Therefore:

$$A = (x + 8) \times 2x = 2x^2 + 16x$$

The possible values for the length and width can be shown in a table, with input x and output A. If the equation was graphed, the possible area values can be seen on the y-axis for given x-values.

Exponential growth and decay can be found in real-world situations. For example, if a piece of notebook paper is folded 25 times, the thickness of the paper can be found. To model this situation, a table can be used. The initial point is one-fold, which yields a thickness of 2 papers. For the second fold, the thickness is 4. Since the thickness doubles each time, the table below shows the thickness for the next few folds. Notice the thickness changes by the same factor each time. Since this change for a constant interval of folds is a factor of 2, the function is exponential. The equation for this is $y = 2^x$. For twenty-five folds, the thickness would be 33,554,432 papers.

x (folds)	y (paper thickness)
0	1
1	2
2	4
3	8
4	16
5	32

One exponential formula that is commonly used is the **interest formula**: $A = Pe^{rt}$. In this formula, interest is compounded continuously. A is the value of the investment after the time, t, in years. P is the initial amount of the investment, r is the interest rate, and e is the constant equal to approximately

2.718. Given an initial amount of \$200 and a time of 3 years, if interest is compounded continuously at a rate of 6%, the total investment value can be found by plugging each value into the formula. The invested value at the end is \$239.44. In more complex problems, the final investment may be given, and the rate may be the unknown. In this case, the formula becomes $239.44 = 200e^{r3}$. Solving for r requires isolating the exponential expression on one side by dividing by 200, yielding the equation $1.20 = e^{r3}$. Taking the natural log of both sides results in $\ln(1.2) = r3$. Using a calculator to evaluate the logarithmic expression, $r = 0.06 = 6\%$.

When working with logarithms and exponential expressions, it is important to remember the relationship between the two. In general, the logarithmic form is $y = log_b x$ for an exponential form $b^y = x$. Logarithms and exponential functions are inverses of each other.

Evaluating Whether a Particular Mathematical Model Can be Used to Describe a Given Set of Conditions

Mathematical functions such as polynomials, rational functions, radical functions, absolute value functions, and linear functions can be utilized to approximate, or model, real-life phenomena. For example, a function can be built that approximates the average amount of snowfall on a given day of the year in Chicago. This example could be as simple as a polynomial. Modeling situations using such functions has limitations; the most significant issue is the error that exists between the exact amount and the approximate amount. Typically, the model will not give exact values as outputs. However, choosing the type of function that provides the best fit of the data will reduce this error. Technology can be used to model situations. For example, given a set of data, the data can be inputted into tools such as graphing calculators or spreadsheet software that output a function with a good fit. Some examples of polynomial modeling are linear, quadratic, and cubic regression.

Linear regression lines are straight lines that calculate a relationship between nonlinear data involving an independent variable and a dependent variable. A regression line is of the form $y = mx + b$, where m is the slope and b is the y-intercept. Both the slope and y-intercept are found using the **Method of Least Squares**, which involves minimizing residuals – the difference between the dependent variable from the data set and the estimated value located on the regression line. The slope represents the rate of change in y as x increases. The y-intercept is the predicted value when the independent variable is equal to 0. Technology, such as a graphing calculator or Microsoft Excel®, can also be utilized to find the equation of this line. In either case, the data points are entered, and a line is "fit" that best represents the shape of the data.

Here is an example of a data set and its regression line:

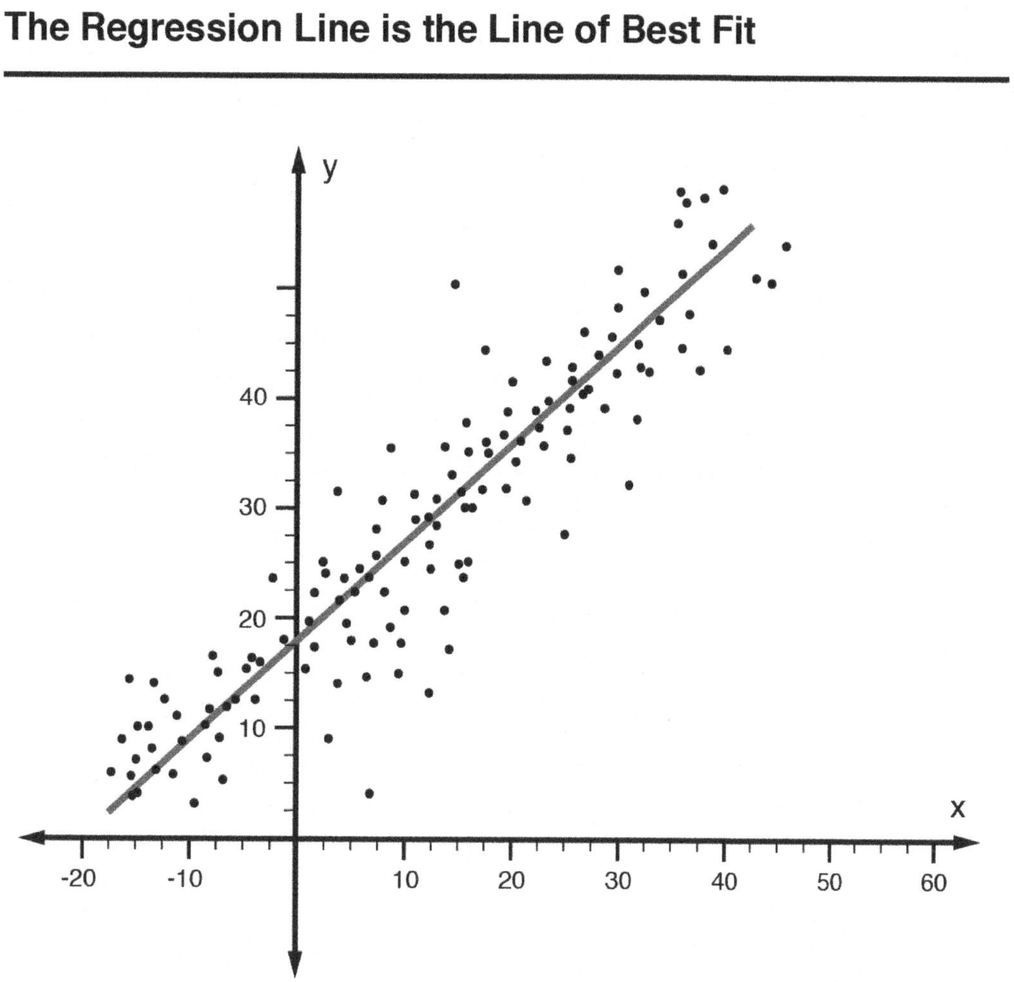

Regression models are highly used for forecasting, and linear regression techniques are the simplest models. If the nonlinear data follows the shape of exponential, logarithmic, or power functions, those types of functions can be used to more accurately model the data rather than lines.

Here is an example of both an exponential regression and a logarithmic regression model:

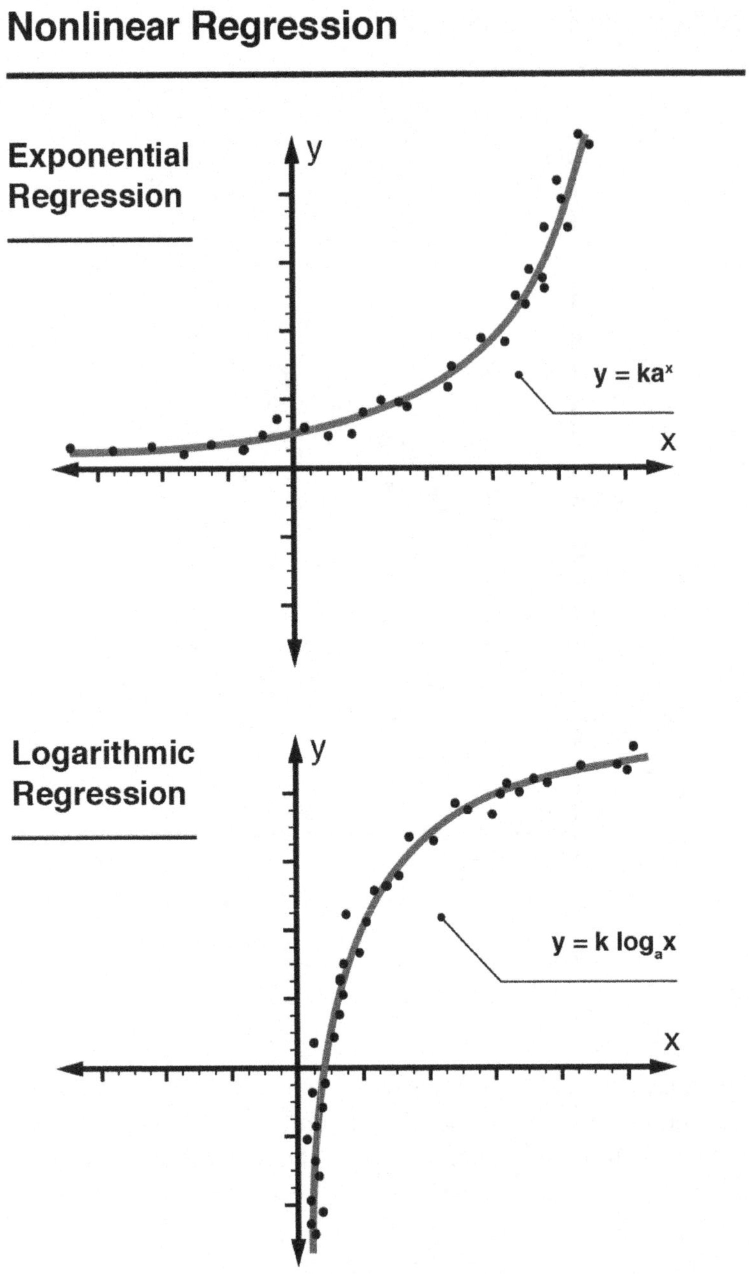

In general, the closer the true data points fall on a regression line or the closer the predicted output (found by using whatever equation is determined to best model a situation) is to the calculated one, the more accurate the model is for the given situation.

Practice Questions

1. The variable *y* is directly proportional to *x*. If $y = 3$ when $x = 5$, then what is *y* when $x = 20$?
   a. 10
   b. 12
   c. 14
   d. 16

2. A line passes through the point (1, 2) and crosses the *y*-axis at $y = 1$. Which of the following is an equation for this line?
   a. $y = 2x$
   b. $y = x + 1$
   c. $x + y = 1$
   d. $y = \frac{x}{2} - 2$

3. An accounting firm charted its income on the following pie graph. If the total income for the year was \$500,000, how much of the income was received from Audit and Taxation Services?

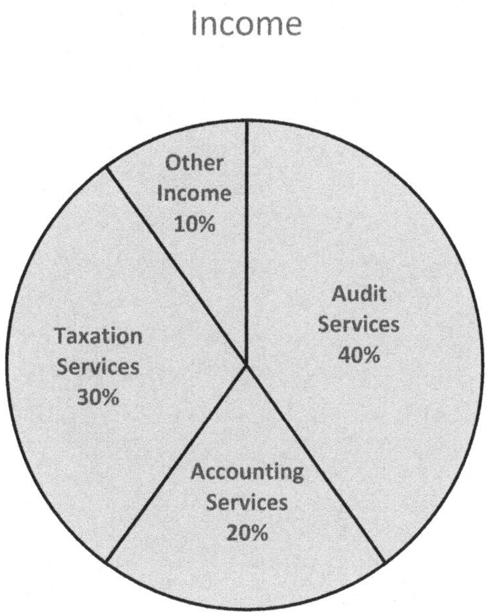

   a. \$200,000
   b. \$350,000
   c. \$150,000
   d. \$300,000

4. $(4x^2y^4)^{\frac{3}{2}}$ can be simplified to which of the following?

a. $8x^3y^6$

b. $4x^{\frac{5}{2}}y$

c. $4xy$

d. $32x^{\frac{7}{2}}y^{\frac{11}{2}}$

5. Gary is driving home to see his parents for Christmas. He travels at a constant speed of 60 miles per hour for a total of 350 miles. How many minutes will it take him to travel home if he takes a break for 10 minutes every 100 miles? Enter your answer in the box below.

Answer Explanations

1. B: To be directly proportional means that $y = mx$. If *x* is changed from 5 to 20, the value of *x* is multiplied by 4. Applying the same rule to the y-value, also multiply the value of *y* by 4. Therefore:

$$y = 12$$

2. B: From the slope-intercept form, $y = mx + b$, it is known that *b* is the *y*-intercept, which is 1. Compute the slope as $\frac{2-1}{1-0} = 1$, so the equation should be $y = x + 1$.

3. B: $350,000: Since the total income is $500,000, then a percentage of that can be found by multiplying the percent of Audit Services as a decimal, or 0.40, by the total of 500,000. This answer is found from the equation:

$$500000 \times 0.4 = 200000$$

The total income from Audit Services is $200,000.

For the income received from Taxation Services, the following equation can be used:

$$500000 \times 0.3 = 150000$$

The total income from Audit Services and Taxation Services is $150{,}000 + 200{,}000 = 350{,}000$.

Another way of approaching the problem is to calculate the easy percentage of 10%, then multiply it by 7, because the total percentage for Audit and Taxation Services was 70%. 10% of 500,000 is 50,000. Then multiplying this number by 7 yields the same income of $350,000.

4. A: Simplify this to:

$$(4x^2y^4)^{\frac{3}{2}} = 4^{\frac{3}{2}}(x^2)^{\frac{3}{2}}(y^4)^{\frac{3}{2}}$$

$$4^{\frac{3}{2}} = (\sqrt{4})^3 = 2^3 = 8$$

For the other, recall that the exponents must be multiplied; this yields:

$$8x^{2\cdot\frac{3}{2}}y^{4\cdot\frac{3}{2}} = 8x^3y^6$$

5. 380 miles. To find the total driving time, the total distance of 350 miles can be divided by the constant speed of 60 miles per hour. This yields a time of 5.8333 hours, which is then rounded. Once the driving time is computed, the break times need to be found. If Gary takes a break for 10 minutes every 100 miles, he will take 3 breaks on his trip. This will yield a total of 30 minutes of break time. Since the answer is needed in minutes, 5.8333 can be converted to minutes by multiplying by 60, giving a driving time of 350 minutes. Adding the break time of 30 minutes to the driving time of 350 minutes gives a total travel time of 380 minutes.

Geometry and Data

Geometry and Measurement

Perimeter, Area, Surface Area, and Volume

Perimeter and Area of Geometric Shapes

Perimeter is the measurement of a distance around something or the sum of all sides of a polygon. Think of perimeter as the length of the boundary, like a fence. In contrast, **area** is the space occupied by a defined enclosure, like a field enclosed by a fence.

When thinking about perimeter, think about walking around the outside of something. When thinking about area, think about the amount of space or **surface area** something takes up.

Square

The perimeter of a square is measured by adding together all of the sides. Since a square has four equal sides, its perimeter can be calculated by multiplying the length of one side by 4. Thus, the formula is $P = 4 \times s$, where s equals one side. For example, the following square has side lengths of 5 meters:

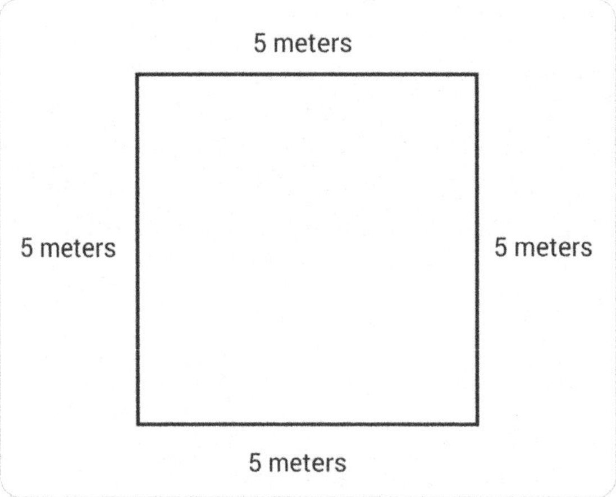

The perimeter is 20 meters because 4 times 5 is 20.

The area of a square is the length of a side squared. For example, if a side of a square is 7 centimeters, then the area is 49 square centimeters. The formula for this example is $A = s^2 = 7^2 = 49$ square centimeters. An example is if the rectangle has a length of 6 inches and a width of 7 inches, then the area is 42 square inches:

$$A = lw = 6(7) = 42 \text{ square inches}$$

Rectangle

Like a square, a rectangle's perimeter is measured by adding together all of the sides. But as the sides are unequal, the formula is different. A rectangle has equal values for its lengths (long sides) and equal values for its widths (short sides), so the perimeter formula for a rectangle is:

$$P = l + l + w + w = 2l + 2w$$

l equals length
w equals width

The area is found by multiplying the length by the width, so the formula is $A = l \times w$.

For example, if the length of a rectangle is 10 inches and the width 8 inches, then the perimeter is 36 inches because:

$$P = 2l + 2w = 2(10) + 2(8) = 20 + 16 = 36 \text{ inches}$$

Triangle

A triangle's perimeter is measured by adding together the three sides, so the formula is $P = a + b + c$, where a, b, and c are the values of the three sides. The area is the product of one-half the base and height so the formula is:

$$A = \frac{1}{2} \times b \times h$$

It can be simplified to:

$$A = \frac{bh}{2}$$

The base is the bottom of the triangle, and the height is the distance from the base to the peak. If a problem asks to calculate the area of a triangle, it will provide the base and height.

For example, if the base of the triangle is 2 feet and the height 4 feet, then the area is 4 square feet. The following equation shows the formula used to calculate the area of the triangle:

$$\text{A} = \frac{1}{2}\text{bh} = \frac{1}{2}(2)(4) = 4 \text{ square feet}$$

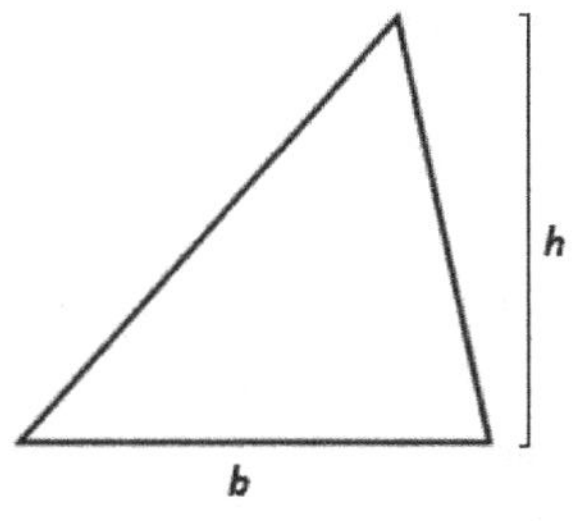

Circle

A circle's perimeter—also known as its circumference—is measured by multiplying the diameter by π.

Diameter is the straight line measured from a point on one side of the circle to a point directly across on the opposite side of the circle.

π is referred to as pi and is equal to 3.14 (with rounding).

So the formula is $\pi \times d$.

This is sometimes expressed by the formula $C = 2 \times \pi \times r$, where r is the radius of the circle. These formulas are equivalent, as the radius equals half of the diameter.

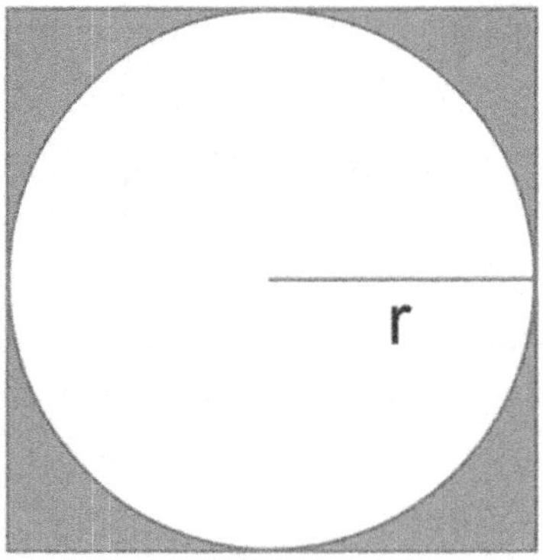

The area of a circle is calculated through the formula $A = \pi \times r^2$. The test will indicate either to leave the answer with π attached or to calculate to the nearest decimal place, which means multiplying by 3.14 for π.

Surface Area and Volume of Geometric Shapes

The area of a two-dimensional figure refers to the number of square units needed to cover the interior region of the figure. This concept is similar to wallpaper covering the flat surface of a wall. For example, if a rectangle has an area of 8 square inches (written 8 in^2), it will take 8 squares, each with sides one inch in length, to cover the interior region of the rectangle. Note that area is measured in square units such as: square feet or ft^2; square yards or yd^2; square miles or mi^2.

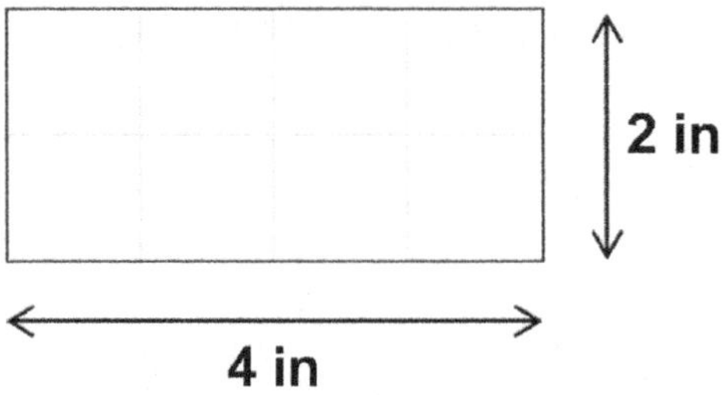

The surface area of a three-dimensional figure refers to the number of square units needed to cover the entire surface of the figure. This concept is similar to using wrapping paper to completely cover the outside of a box. For example, if a triangular pyramid has a surface area of 17 square inches (written $17in^2$), it will take 17 squares, each with sides one inch in length, to cover the entire surface of the pyramid. Surface area is also measured in square units.

Many three-dimensional figures (solid figures) can be represented by nets consisting of rectangles and triangles. The surface area of such solids can be determined by adding the areas of each of its faces and bases. Finding the surface area using this method requires calculating the areas of rectangles and triangles. To find the area (*A*) of a rectangle, the length (*l*) is multiplied by the width $(w) \rightarrow A = l \times w$. The area of a rectangle with a length of 8cm and a width of 4cm is calculated: $A = (8cm) \times (4cm) \rightarrow A = 32cm^2$.

To calculate the area (A) of a triangle, the product of $\frac{1}{2}$, the base (b), and the height (h) is found → $A = \frac{1}{2} \times b \times h$. Note that the height of a triangle is measured from the base to the vertex opposite of it forming a right angle with the base. The area of a triangle with a base of 11cm and a height of 6cm is calculated: $A = \frac{1}{2} \times (11cm) \times (6cm) \rightarrow A = 33cm^2$.

Consider the following triangular prism, which is represented by a net consisting of two triangles and three rectangles.

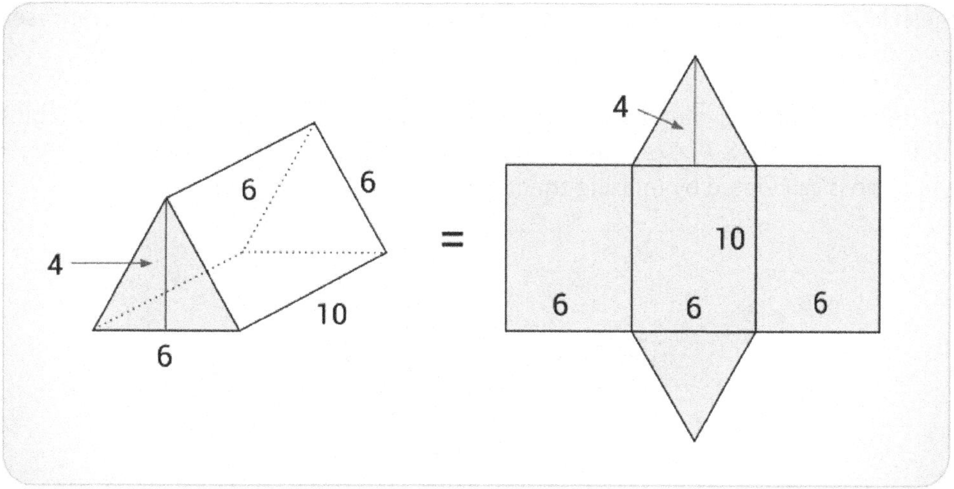

The surface area of the prism can be determined by adding the areas of each of its faces and bases. The surface area (*SA*) = area of triangle + area of triangle + area of rectangle + area of rectangle + area of rectangle.

$$SA = \left(\frac{1}{2} \times b \times h\right) + \left(\frac{1}{2} \times b \times h\right) + (l \times w) + (l \times w) + (l \times w)$$

$$SA = \left(\frac{1}{2} \times 6 \times 4\right) + \left(\frac{1}{2} \times 6 \times 4\right) + (6 \times 10) + (6 \times 10) + (6 \times 10)$$

$$SA = (12) + (12) + (60) + (60) + (60)$$

$$SA = 204 \textit{ square units}$$

Volume is the measurement of how much space an object occupies, like how much space is in the cube. Volume is useful in determining the space within a certain three-dimensional object. Volume can be calculated for a cube, rectangular prism, cylinder, pyramid, cone, and sphere. By knowing specific dimensions of the objects, the volume of the object is computed with these figures. The units for the volumes of solids can include cubic centimeters, cubic meters, cubic inches, and cubic feet.

Cube

The cube is the simplest figure for which volume can be determined because all dimensions in a cube are equal. In the following example, the length, width, and height of the cube are all represented by the variable *a* because these measurements are equal lengths.

The volume of any rectangular, three-dimensional object is found by multiplying its length by its width by its height. In the case of a cube, the length, width, and height are all equal lengths, represented by the variable *a*. Therefore, the equation used to calculate the volume is $(a \times a \times a)$ or a^3. In a real-world example of this situation, if the length of a side of the cube is 3 centimeters, the volume is calculated by utilizing the formula $(3 \times 3 \times 3) = 9$ cm³.

Rectangular Prism

The dimensions of a rectangular prism are not necessarily equal as those of a cube. Therefore, the formula for a rectangular prism recognizes that the dimensions vary and use different variables to represent these lengths. The length, width, and height of a rectangular prism can be represented with the variables *a, b*, and *c*.

The equation used to calculate volume is length times width times height. Using the variables in the diagram above, this means $a \times b \times c$. In a real-world application of this situation, if *a*=2 cm, *b*=3 cm, and *c*=4 cm, the volume is calculated by utilizing the formula $3 \times 4 \times 5 = 60$ cm³.

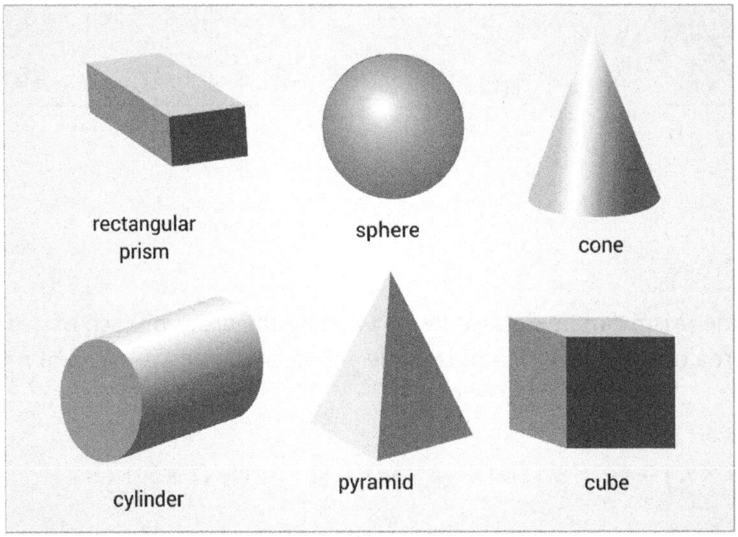

Cylinder

Discovering a cylinder's volume requires the measurement of the cylinder's base, length of the radius, and height. The height of the cylinder can be represented with variable *h*, and the radius can be represented with variable *r*.

The formula to find the volume of a cylinder is $\pi r^2 h$. Notice that πr^2 is the formula for the area of a circle. This is because the base of the cylinder is a circle. To calculate the volume of a cylinder, the slices of circles needed to build the entire height of the cylinder are added together. For example, if the radius is 5 feet and the height of the cylinder is 10 feet, the cylinder's volume is calculated by using the following equation: $\pi 5^2 \times 10$. Substituting 3.14 for π, the volume is 785.4 ft³.

Pyramid

To calculate the volume of a pyramid, the area of the base of the pyramid is multiplied by the pyramid's height by $\frac{1}{3}$. The area of the base of the pyramid is found by multiplying the base length by the base width.

Therefore, the formula to calculate a pyramid's volume is $(L \times W \times H) \div 3$.

Cone

The formula to calculate the volume of a circular cone is similar to the formula for the volume of a pyramid. The primary difference in determining the area of a cone is that a circle serves as the base of a cone. Therefore, the area of a circle is used for the cone's base.

The variable *r* represents the radius, and the variable *h* represents the height of the cone. The formula used to calculate the volume of a cone is $\frac{1}{3}\pi r^2 h$. Essentially, the area of the base of the cone is multiplied by the cone's height. In a real-life example where the radius of a cone is 2 meters and the height of a cone is 5 meters, the volume of the cone is calculated by utilizing the formula $\frac{1}{3}\pi 2^2 \times 5$. After substituting 3.14 for π, the volume is 20.93 ft^3.

Sphere

The volume of a sphere uses π due to its circular shape.

The length of the radius, *r*, is the only variable needed to determine the sphere's volume. The formula to calculate the volume of a sphere is $\frac{4}{3}\pi r^3$. Therefore, if the radius of a sphere is 8 centimeters, the volume of the sphere is calculated by utilizing the formula $\frac{4}{3}\pi(8)^3$ =2,143 cm^3.

Using Two-Dimensional Representations of Three-Dimensional Objects to Visualize and Solve Problems

Three-dimensional objects can be simplified into related two-dimensional shapes to solve problems. This simplification can make problem-solving a much easier experience. An isometric representation of a three-dimensional object can be completed so that important properties (e.g., shape, relationships of faces and surfaces) are noted. Edges and vertices can be translated into two-dimensional objects as well.

For example, below is a three-dimensional object that's been partitioned into two-dimensional representations of its faces:

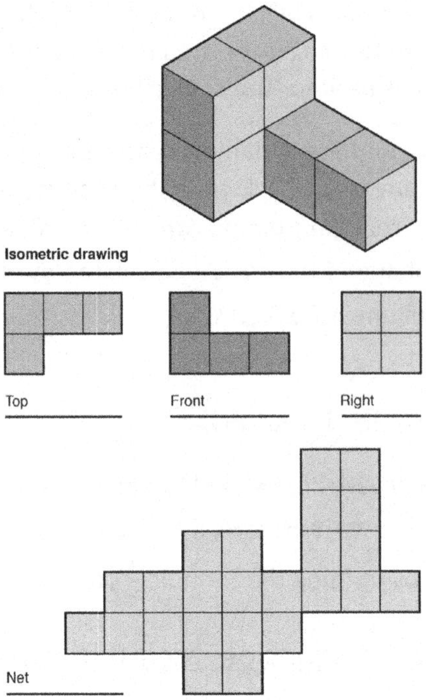

The net represents the sum of the three different faces. Depending on the problem, using a smaller portion of the given shape may be helpful, by simplifying the steps necessary to solve.

Many objects in the real world consist of three-dimensional shapes such as prisms, cylinders, and spheres. Surface area problems involve quantifying the outside area of such a three-dimensional object, and volume problems involve quantifying how much space the object takes up. As mentioned, surface area of a prism is the sum of the areas, which is simplified into $SA = 2A + Bh$, where *A* is the area of the base, *B* is the perimeter of the base, and *h* is the height of the prism. The volume of the same prism is $V = Ah$. The surface area of a cylinder is equal to the sum of the areas of each end and the side, which is $SA = 2\pi rh + 2\pi r^2$, and its volume is $V = \pi r^2 h$. Finally, the surface area of a sphere is $SA = 4\pi r^2$, and its volume is $V = \frac{4}{3}\pi r^3$.

An example when one of these formulas should be used would be when calculating how much paint is needed for the outside of a house. In this scenario, surface area must be used. The sum of all individual areas of each side of the house must be found. Also, when calculating how much water a cylindrical tank can hold, a volume formula is used. Therefore, the amount of water that a cylindrical tank that is 8 feet tall with a radius of 3 feet is $\pi \times 3^2 \times 8 = 226.1$ cubic feet.

Similarity and Congruence

Two figures are **congruent** if there is a rigid motion that can map one figure onto the other. Therefore, all pairs of sides and angles within the image and pre-image must be congruent. For example, in triangles, each pair of the three sides and three angles must be congruent. Similarly, in two four-sided figures, each pair of the four sides and four angles must be congruent.

Two figures are **similar** if there is a combination of translations, reflections, rotations, and dilations, which maps one figure onto the other. The difference between congruence and similarity is that dilation can be used in similarity. Therefore, side lengths between each shape can differ. However, angle measure must be preserved within this definition. If two polygons differ in size so that the lengths of corresponding line segments differ by the same factor, but corresponding angles have the same measurement, they are similar.

Essentially, two figures are **congruent** if they have the same shape and same size. The two figures could have been rotated, reflected, or translated. Two figures are **similar** if they have been rotated, reflected, translated, and resized. Angle measure is preserved in similar figures. Both angle and side length are preserved in congruent figures.

In **similar figures**, if the ratio of two corresponding sides is known, then that ratio, or **scale factor**, holds true for all of the dimensions of the new figure.

Using Similarity and Congruence to Solve Problems with Two-Dimensional and Three-Dimensional Figures

Similarity and congruence can be used to solve problems. Suppose that Lara is 5 feet tall and is standing 30 feet from the base of a light pole, and her shadow is 6 feet long. How high is the light on the pole?

To figure this, it helps to make a sketch of the situation:

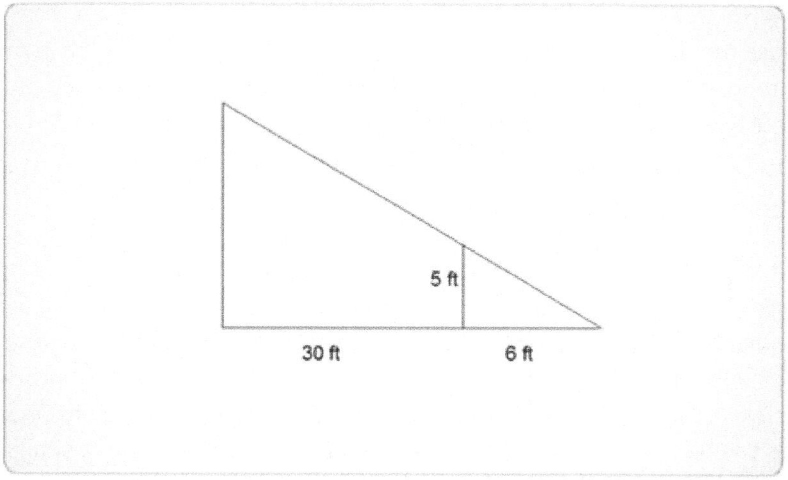

The light pole is the left side of the triangle. Lara is the 5-foot vertical line. Notice that there are two right triangles here, and that they have all the same angles as one another. Therefore, they form similar triangles. So, the ratio of proportionality between them must be determined.

The bases of these triangles are known. The small triangle, formed by Lara and her shadow, has a base of 6 feet. The large triangle formed by the light pole along with the line from the base of the pole out to the end of Lara's shadow is $30 + 6 = 36$ feet long. So, the ratio of the big triangle to the little triangle will be $\frac{36}{6} = 6$. The height of the little triangle is 5 feet. Therefore, the height of the big triangle will be $6 \times 5 = 30$ feet, meaning that the light is 30 feet up the pole.

Notice that the perimeter of a figure changes by the ratio of proportionality between two similar figures, but the area changes by the square of the ratio. This is because if the length of one side is doubled, the area is quadrupled.

As an example, suppose two rectangles are similar, but the edges of the second rectangle are three times longer than the edges of the first rectangle. The area of the first rectangle is 10 square inches. How much more area does the second rectangle have than the first? To answer this, note that the area of the second rectangle is $3^2 = 9$ times the area of the first rectangle, which is 10 square inches. Therefore, the area of the second rectangle is going to be $9 \times 10 = 90$ square inches. This means it has $90 - 10 = 80$ square inches more area than the first rectangle.

As a second example, suppose *X* and *Y* are similar right triangles. The hypotenuse of *X* is 4 inches. The area of *Y* is $\frac{1}{4}$ the area of *X*. What is the hypotenuse of *Y*? First, realize the area has changed by a factor of $\frac{1}{4}$. The area changes by a factor that is the *square* of the ratio of changes in lengths, so the ratio of the lengths is the square root of the ratio of areas. That means that the ratio of lengths must be is $\sqrt{\frac{1}{4}} = \frac{1}{2}$, and the hypotenuse of *Y* must be $\frac{1}{2} \times 4 = 2$ inches.

Volumes between similar solids change like the cube of the change in the lengths of their edges. Likewise, if the ratio of the volumes between similar solids is known, the ratio between their lengths is known by finding the cube root of the ratio of their volumes.

For example, suppose there are two similar rectangular pyramids *X* and *Y*. The base of *X* is 1 inch by 2 inches, and the volume of *X* is 8 inches. The volume of *Y* is 64 inches. What are the dimensions of the base of *Y*? To answer this, first find the ratio of the volume of *Y* to the volume of *X*. This will be given by $\frac{64}{8} = 8$. Now the ratio of lengths is the cube root of the ratio of volumes, or $\sqrt[3]{8} = 2$. So, the dimensions of the base of *Y* must be 2 inches by 4 inches.

The criteria needed to prove triangles are congruent involves both angle and side congruence. Both pairs of related angles and sides need to be of the same measurement to use congruence in a proof. The criteria to prove similarity in triangles involves proportionality of side lengths. Angles must be congruent in similar triangles; however, corresponding side lengths only need to be a constant multiple of each other. Once similarity is established, it can be used in proofs as well. Relationships in geometric figures other than triangles can be proven using triangle congruence and similarity. If a similar or congruent triangle can be found within another type of geometric figure, their criteria can be used to prove a relationship about a given formula. For instance, a rectangle can be broken up into two congruent triangles.

If two angles of one triangle are congruent with two angles of a second triangle, the triangles are similar. This is because, within any triangle, the sum of the angle measurements is 180 degrees. Therefore, if two are congruent, the third angle must also be congruent because their measurements are equal. Three congruent pairs of angles mean that the triangles are similar.

There are five theorems to show that triangles are congruent when it's unknown whether each pair of angles and sides are congruent. Each theorem is a shortcut that involves different combinations of sides and angles that must be true for the two triangles to be congruent. For example, **side-side-side (SSS)** states that if all sides are equal, the triangles are congruent. **Side-angle-side (SAS)** states that if two pairs of sides are equal and the included angles are congruent, then the triangles are congruent.

Similarly, **angle-side-angle (ASA)** states that if two pairs of angles are congruent and the included side lengths are equal, the triangles are similar. **Angle-angle-side (AAS***)* states that two triangles are congruent if they have two pairs of congruent angles and a pair of corresponding equal side lengths that aren't included. Finally, **hypotenuse-leg (HL)** states that if two right triangles have equal hypotenuses and an equal pair of shorter sides, then the triangles are congruent. An important item to note is that angle-angle-angle *(AAA)* is not enough information to have congruence. It's important to understand why these rules work by using rigid motions to show congruence between the triangles with the given properties. For example, three reflections are needed to show why *SAS* follows from the definition of congruence.

Properties of Lines and Angles

Parallel, Perpendicular, and Intersecting Lines

Two lines are parallel if they have the same slope and a different intercept. Two lines are **perpendicular** if the product of their slope equals -1. Parallel lines never intersect unless they are the same line, and perpendicular lines intersect at a right angle. If two lines aren't parallel, they must intersect at one point. If lines do cross, they're labeled as **intersecting lines** because they "intersect" at one point. If they intersect at more than one point, they're the same line. Determining equations of lines based on properties of parallel and perpendicular lines appears in word problems. To find an equation of a line, both the slope and a point the line goes through are necessary. Therefore, if an equation of a line is needed that's parallel to a given line and runs through a specified point, the slope of the given line and the point are plugged into the point-slope form of an equation of a line. Secondly, if an equation of a line is needed that's perpendicular to a given line running through a specified point, the negative reciprocal of the slope of the given line and the point are plugged into the **point-slope form**. Also, if the point of intersection of two lines is known, that point will be used to solve the set of equations. Therefore, to solve a system of equations, the point of intersection must be found. If a set of two equations with two unknown variables has no solution, the lines are parallel.

The **Parallel Postulate** states that if two parallel lines are cut by a transversal, then the corresponding angles are equal. Here is a picture that highlights this postulate:

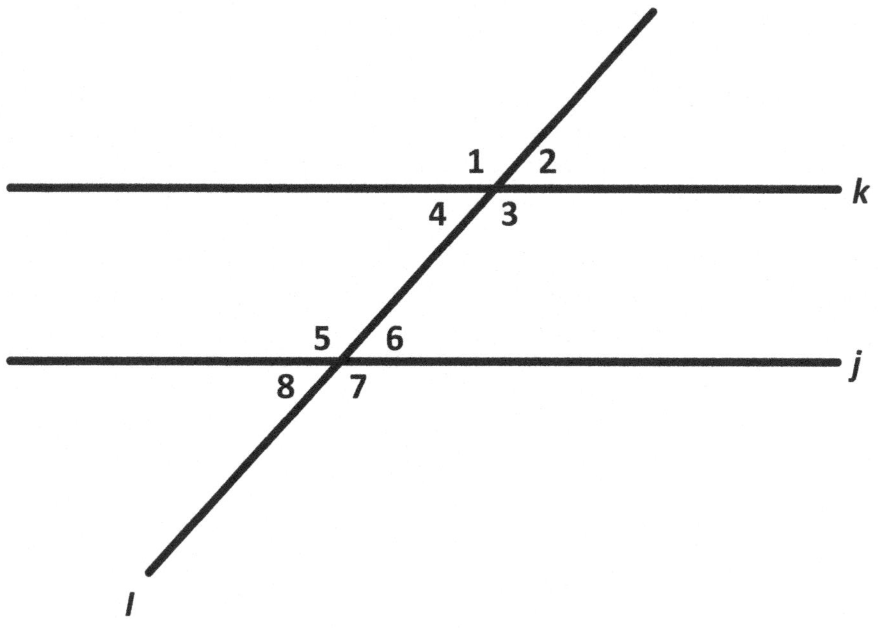

Because lines *k* and *i* are parallel, when cut by transversal *l*, angles 1 and 5 are equal, angles 2 and 6 are equal, angles 4 and 8 are equal, and angles 3 and 7 are equal. Note that angles 1 and 2, 3 and 4, 5 and 6, and 7 and 8 add up to 180 degrees.

This statement is equivalent to the **Alternate Interior Angle Theorem**, which states that when two parallel lines are cut by a transversal, the resultant interior angles are congruent. In the picture above, angles 3 and 5 are congruent, and angles 4 and 6 are congruent.

The Parallel Postulate or the Alternate Interior Angle Theorem can be used to find the missing angles in the following picture:

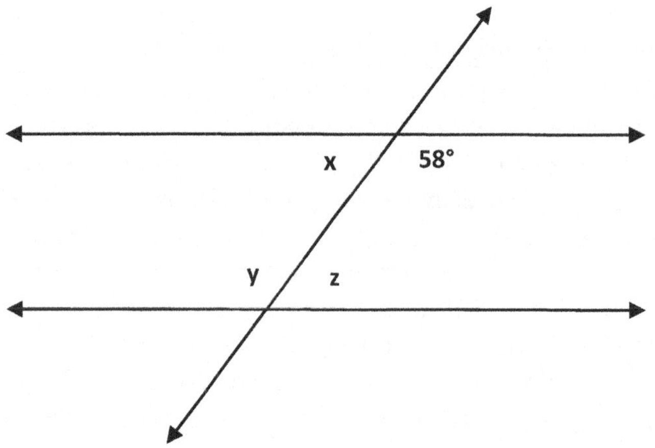

Assuming that the lines are parallel, angle x is found to be 122 degrees. Angle x and the 58-degree angle add up to 180 degrees. The Alternate Interior Angle Theorem states that angle y is equal to 58 degrees. Also, angles y and z add up to 180 degrees, so angle z is 122 degrees. Note that angles x and z are also alternate interior angles, so their equivalence can be used to find angle z as well.

An equivalent statement to the Parallel Postulate is that the sum of all angles in a triangle is 180 degrees. Therefore, given any triangle, if two angles are known, the third can be found accordingly.

Applying Angle Relationships to Solve Problems

An angle consists of two rays that have a common endpoint. This common endpoint is called the vertex of the angle. The two rays can be called sides of the angle. The angle below has a vertex at point *B* and the sides consist of ray *BA* and ray *BC*. An angle can be named in three ways:

1. Using the vertex and a point from each side, with the vertex letter in the middle.
2. Using only the vertex. This can only be used if it is the only angle with that vertex.
3. Using a number that is written inside the angle.

The angle below can be written $\angle ABC$ (read angle *ABC*), $\angle CBA$, $\angle B$, or $\angle 1$.

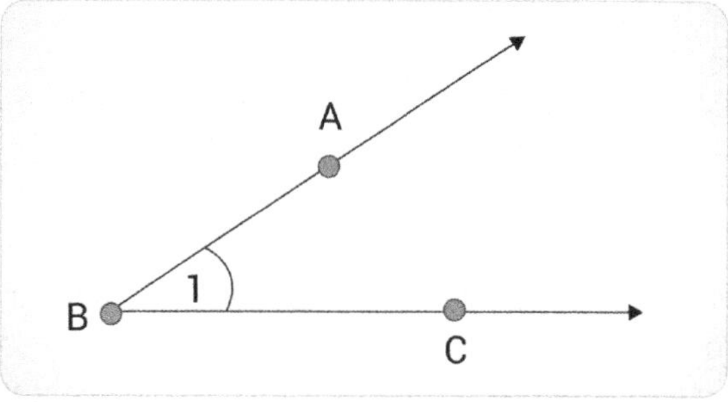

An angle divides a plane, or flat surface, into three parts: the angle itself, the interior (inside) of the angle, and the exterior (outside) of the angle. The figure below shows point *M* on the interior of the angle and point *N* on the exterior of the angle.

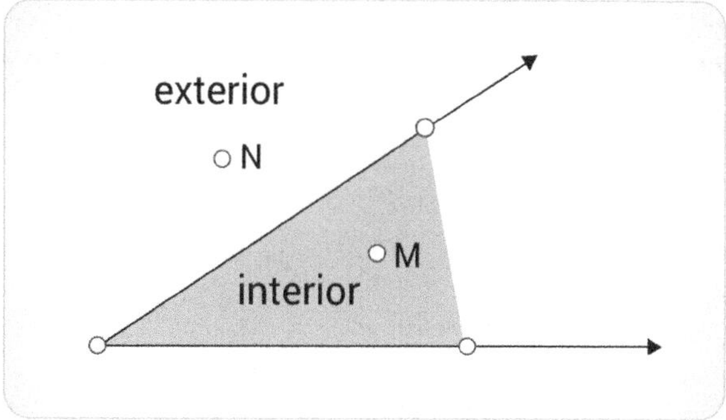

Angles can be measured in units called degrees, with the symbol °. The degree measure of an angle is between 0° and 180°, is a measure of rotation, and can be obtained by using a protractor.

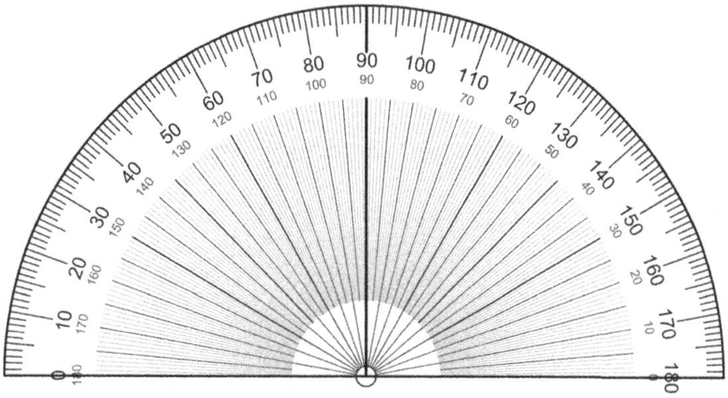

A straight angle (or simply a line) measures exactly 180°. A right angle's sides meet at the vertex to create a square corner. A right angle measures exactly 90° and is typically indicated by a box drawn in the interior of the angle. An acute angle has an interior that is narrower than a right angle. The measure of an acute angle is any value less than 90° and greater than 0°. For example, 89.9°, 47°, 12°, and 1°. An obtuse angle has an interior that is wider than a right angle. The measure of an obtuse angle is any value greater than 90° but less than 180°. For example, 90.1°, 110°, 150°, and 179.9°. Any two angles that sum up to 90 degrees are known as *complementary angles.*

- Acute angles: Less than 90°
- Obtuse angles: Greater than 90°
- Right angles: 90°
- Straight angles: 180°

To determine angle measures for adjacent angles, angles that share a common side and vertex, at least one of the angles must be known. Other information that is necessary to determine such measures include that there are 90° in a right angle, and there are 180° in a straight line. Therefore, if two adjacent angles form a right angle, they will add up to 90°, and if two adjacent angles form a straight line, they add up to 180°.

If the measurement of one of the adjacent angles is known, the other can be found by subtracting the known angle from the total number of degrees.

For example, given the following situation, if angle *a* measures 55°, find the measure of unknown angle *b*:

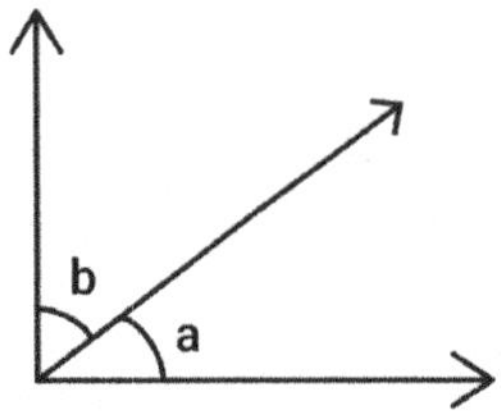

To solve this simply subtract the known angle measure from 90°.

$$90° - 55° = 35°$$

The measure of *b* = 35°.

Given the following situation, if angle 1 measures 45°, find the measure of the unknown angle 2:

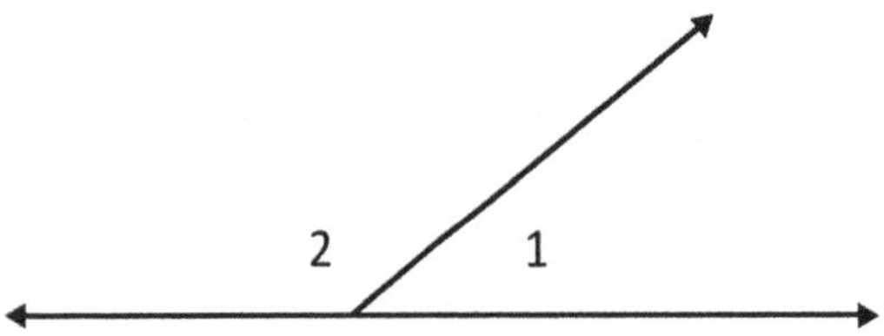

To solve this, simply subtract the known angle measure from 180°.

$$180° - 45° = 135°$$

The measure of angle 2 = 135°.

In the case that more than two angles are given, use the same method of subtracting the known angles from the total measure.

For example, given the following situation, if angle $y = 40^\circ$, and angle $z = 25^\circ$, find unknown angle x.

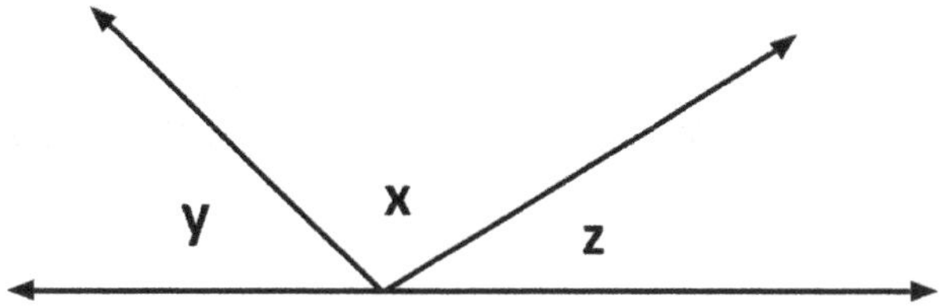

Subtract the known angles from 180°.

$$180^\circ - 40^\circ - 25^\circ = 115^\circ$$

The measure of angle $x = 115^\circ$.

Properties of Triangles

Relationships Between Sides and Angles

The **Exterior Angle Theorem** states that the measure of an exterior angle of a triangle is equal to the sum of the measures of the two non-adjacent interior angles inside the triangle. It can be shown here:

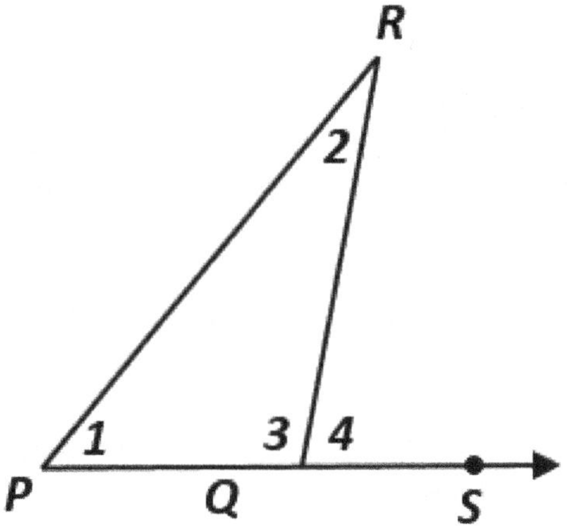

The measure of angle 4 is equal to the sum of the measures of angles 1 and 2. Also, the Triangle Inequality Theorem states that for any triangle, the sum of the lengths of two sides is greater than or equal to the length of the third side. For example, in the picture above, $PQ + QR \geq PR$.

Within similar triangles, corresponding sides are proportional, and angles are congruent. In addition, within similar triangles, the ratio of the side lengths is the same. This property is true even if side lengths are different. Within right triangles, trigonometric ratios can be defined for the acute angle within the triangle. The functions are defined through ratios in a right triangle. **Sine** of acute angle, A, is opposite over hypotenuse, **cosine** is adjacent over hypotenuse, and **tangent** is opposite over adjacent. Note that expanding or shrinking the triangle won't change the ratios. However, changing the angle measurements will alter the calculations.

Within a right triangle, two complementary angles exist because the third angle is always 90 degrees. In this scenario, the *sine* of one of the complementary angles is equal to the cosine of the other angle. The opposite is also true. This relationship exists because sine and cosine will be calculated as the ratios of the same side lengths.

The **Pythagorean theorem** is an important relationship between the three sides of a right triangle. It states that the square of the side opposite the right triangle, known as the **hypotenuse** (denoted as c^2), is equal to the sum of the squares of the other two sides ($a^2 + b^2$). Thus, $a^2 + b^2 = c^2$. As an example of the theorem, suppose that Shirley has a rectangular field that is 5 feet wide and 12 feet long, and she wants to split it in half using a fence that goes from one corner to the opposite corner. How long will this fence need to be? To figure this out, note that this makes the field into two right triangles, whose hypotenuse will be the fence dividing it in half. Therefore, the fence length will be given by $\sqrt{5^2 + 12^2} = \sqrt{169} = 13$ feet long.

The converse of the Pythagorean Theorem is another important property of triangles, and it states that if the square of one side of a triangle is equal to the sum of the squares of the other two sides, then the triangle must have a 90-degree angle and therefore is a right triangle.

Both the trigonometric functions and the Pythagorean theorem can be used in problems that involve finding either a missing side or a missing angle of a right triangle. To do so, one must look to see what sides and angles are given and select the correct relationship that will help find the missing value. These relationships can also be used to solve application problems involving right triangles. Often, it's helpful to draw a figure to represent the problem to see what's missing.

Medians, Midpoints, and Altitudes

A **median** of a triangle is the line drawn from a vertex to the midpoint on the opposite side. A triangle has three medians, and their point of intersection is known as the **centroid**. An **altitude** is a line drawn from a vertex perpendicular to the opposite side. A triangle has three altitudes, and their point of intersection is known as the **orthocenter**. An altitude can actually exist outside, inside, or on the triangle depending on the placement of the vertex. Many problems involve these definitions. For example, given one endpoint of a line segment and the midpoint, the other endpoint can be determined by using the midpoint formula. In addition, area problems heavily depend on these definitions. For example, it can be proven that the median of a triangle divides it into two regions of equal areas. The actual formula for the area of a triangle depends on its altitude.

Special Triangles

An **isosceles triangle** contains at least two equal sides. Therefore, it must also contain two equal angles and, subsequently, contain two medians of the same length. An isosceles triangle can also be labelled as an **equilateral triangle** (which contains three equal sides and three equal angles) when it meets these conditions. In an equilateral triangle, the measure of each angle is always 60 degrees. Also within an equilateral triangle, the medians are of the same length. A **scalene triangle** can never be an equilateral or an isosceles triangle because it contains no equal sides and no equal angles. Also, medians in a scalene triangle can't have the same length. However, a **right triangle**, which is a triangle containing a 90-degree angle, can be a scalene triangle. There are two types of special right triangles. The **30-60-90 right triangle** has angle measurements of 30 degrees, 60 degrees, and 90 degrees. Because of the nature of this triangle, and through the use of the Pythagorean theorem, the side lengths have a special relationship. If x is the length opposite the 30-degree angle, the length opposite the 60-degree angle is $\sqrt{3}x$, and the hypotenuse has length $2x$. The **45-45-90 right triangle** is also special as it contains two angle measurements of 45 degrees. It can be proven that, if x is the length of the two equal sides, the hypotenuse is $x\sqrt{2}$. The properties of all of these special triangles are extremely useful in determining both side lengths and angle measurements in problems where some of these quantities are given and some are not.

Properties of Quadrilaterals and Other Polygons

Geometric Properties of Various Quadrilaterals

A special quadrilateral is one in which both pairs of opposite sides are parallel. This type of quadrilateral is known as a **parallelogram**. A parallelogram has six important properties:

- Opposite sides are congruent.
- Opposite angles are congruent.
- Within a parallelogram, consecutive angles are supplementary, so their measurements total 180 degrees.
- If one angle is a right angle, all of them have to be right angles.
- The diagonals of the angles bisect each other.
- These diagonals form two congruent triangles.

A parallelogram with four congruent sides is a **rhombus**. A quadrilateral containing only one set of parallel sides is known as a **trapezoid**. The parallel sides are known as **bases**, and the other two sides are known as **legs**. If the legs are congruent, the trapezoid can be labelled an **isosceles trapezoid**. An important property of a trapezoid is that their **diagonals** are congruent. Also, the **median** of a trapezoid is parallel to the bases, and its length is equal to half of the sum of the base lengths.

Rectangles, squares, and rhombuses are **polygons** with four sides. By definition, all rectangles are parallelograms, but only some rectangles are squares. However, some parallelograms are rectangles. Also, it's true that all squares are rectangles, and some rhombuses are squares. There are no rectangles, squares, or rhombuses that are trapezoids though, because they have more than one set of parallel sides.

If a quadrilateral is inscribed in a circle, the sum of its opposite angles is 180 degrees. Consider the quadrilateral ABCD centered at the point O:

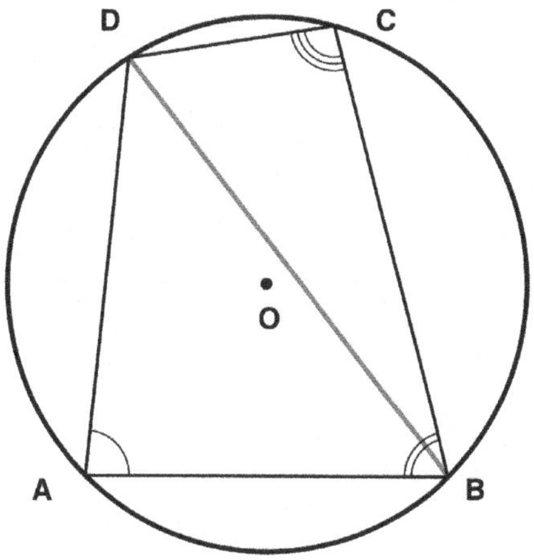

Each of the four line segments within the quadrilateral is a **chord** of the circle. Consider the diagonal DB. Angle DAB is an inscribed angle leaning on the arc DCB. Therefore, angle DAB is half the measure of the arc DCB. Conversely, angle DCB is an inscribed angle leaning on the arc DAB. Therefore, angle DCB is half the measure of the arc DAB. The sum of arcs DCB and DAB is 360 degrees because they make up the entire circle. Therefore, the sum of angles DAB and DCB equals half of 360 degrees, which is 180 degrees.

Relationships Among Quadrilaterals

Rectangles, squares, and rhombuses are **polygons** with four sides. By definition, all rectangles are parallelograms, but only some rectangles are squares. However, some parallelograms are rectangles. Also, it's true that all squares are rectangles, and some rhombuses are squares. There are no rectangles, squares, or rhombuses that are trapezoids though, because they have more than one set of parallel sides.

Quadrilaterals can be further classified according to their sides and angles. A quadrilateral with exactly one pair of parallel sides is called a trapezoid. A quadrilateral that shows both pairs of opposite sides parallel is a parallelogram. Parallelograms include rhombuses, rectangles, and squares. A rhombus has four equal sides. A rectangle has four equal angles (90° each). A square has four 90° angles and four equal sides. Therefore, a square is both a rhombus and a rectangle.

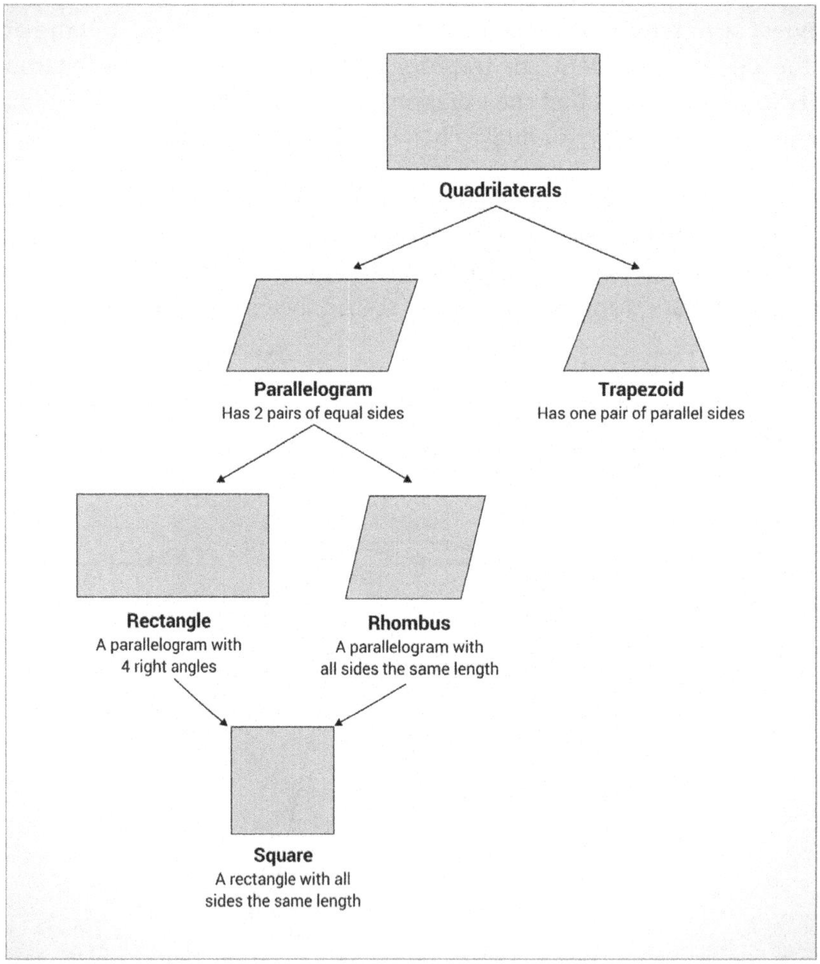

Angles and Diagonals

Diagonals are lines (excluding sides) that connect two vertices within a polygon. **Mutually bisecting diagonals** intersect at their midpoints. Parallelograms, rectangles, squares, and rhombuses have mutually bisecting diagonals. However, trapezoids don't have such lines. **Perpendicular diagonals** occur when they form four right triangles at their point of intersection. Squares and rhombuses have perpendicular diagonals, but trapezoids, rectangles, and parallelograms do not. Finally, **perpendicular bisecting diagonals** (also known as **perpendicular bisectors**) form four right triangles at their point of intersection, but this intersection is also the midpoint of the two lines. Both rhombuses and squares have perpendicular bisecting angles, but trapezoids, rectangles, and parallelograms do not. Knowing these definitions can help tremendously in problems that involve both angles and diagonals.

Polygons with More than Four Sides

A **pentagon** is a five-sided figure. A six-sided shape is a **hexagon**. A seven-sided figure is classified as a **heptagon**, and an eight-sided figure is called an **octagon**. An important characteristic is whether a polygon is regular or irregular. If it's **regular,** the side lengths and angle measurements are all equal. An **irregular** polygon has unequal side lengths and angle measurements. Mathematical problems involving polygons with more than four sides usually involve side length and angle measurements. The sum of all internal angles in a polygon equals $180(n-2)$ degrees, where n is the number of sides. Therefore, the total of all internal angles in a pentagon is 540 degrees because there are five sides so $180(5-2)$ = 540 degrees. Unfortunately, area formulas don't exist for polygons with more than four sides. However, their shapes can be split up into triangles, and the formula for area of a triangle can be applied and totaled to obtain the area for the entire figure.

Properties of Circles

Solving Problems Involving Circumference and Area of a Circle

As mentioned, a circle's perimeter—also known as its **circumference**—is measured by multiplying the diameter by π (approximately, 3.14). Diameter is the straight line measured from a point on one side of the circle to a point directly across on the opposite side of the circle. Circumference can also be calculated with the formula $C = 2 \times \pi \times r$, where r is the radius of the circle. These formulas are equivalent, as the radius equals half of the diameter.

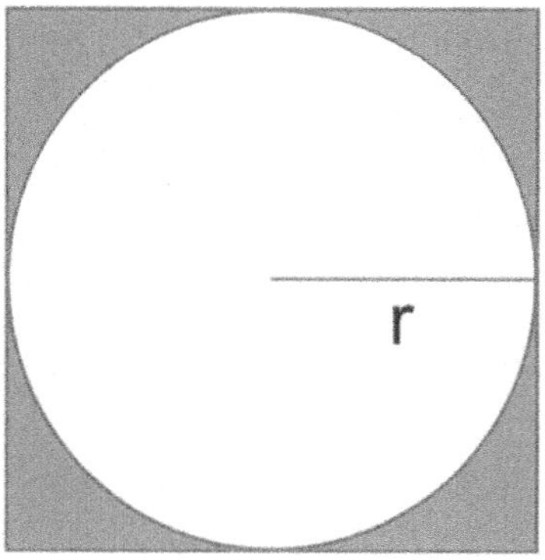

The equation used to find the area of a circle is $A = \pi r^2$. For example, if a circle has a radius of 5 centimeters, the area is computed by substituting 5 for the radius: $(5)^2$.

Consider the following problem: A pizzeria owner regularly creates jumbo pizzas, each with a radius of 9 inches. She is mathematically inclined, and wants to know the area of the pizza to purchase the correct boxes and know how much she is feeding her customers. What is the area of the circle, in terms of π, with a radius of 9 inches?

To solve this problem, the formula for the area of the circle (πr^2). The radius of the pizza is 9 inches so 9 squared is 81. Therefore, the area of the pizza will be 81 π in^2.

Solving Problems Involving Diameter or Radius of a Circle

As mentioned, the **diameter** of a circle is the straight line measured from one end to the direct opposite end of the circle, while the radius is half that distance. Essentially, the **radius** of a circle is the distance from the center of the circle to any point on the circle. For example, if the diameter of a circle is 10 inches, the radius is 5 inches.

Solving Basic Problems Involving Central Angles, Tangents, Arcs, and Sectors

To find any fractional area of a circle, the formula $A = \frac{C}{360}\pi r^2$ can be used, where *C* is the number of degrees of the central angle of the sector. The area of a circle can also be found by using the arc length rather than the degree of the sector. This formula is $A = rs^2$, where *s* is the arc length and *r* is the radius of the circle.

A **chord** of a circle is a straight line formed when its endpoints are allowed to be any two points on the circle. If the radius and the value of the angle subtended at the center by the chord is known, the formula to find the chord length is: $C = 2 \times radius \times \sin\frac{angle}{2}$. Remember that this formula is based on half the length of the chord, so the radius is doubled to determine the full length of the chord.

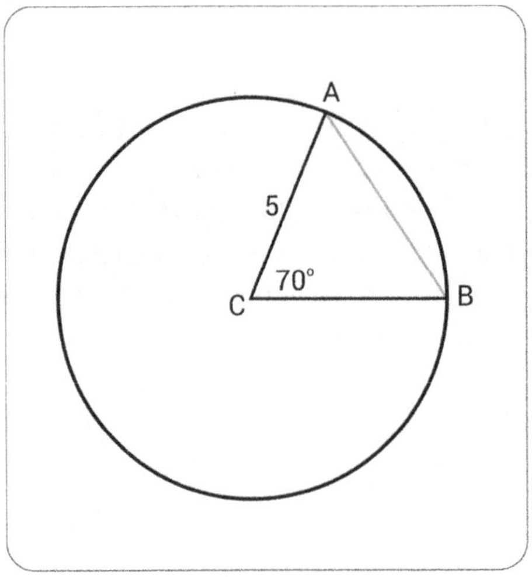

For example, the radius in the diagram above is 5 and the angle is 70 degrees. To find the chord length, plug in the values for the radius and angle to obtain the answer of 5.7.

$$5 \times \sin\frac{70}{2} = 2.87 \times 2 = 5.7$$

Many angles exist within a circle. A **central angle** is formed by using two radii as its rays and the center of the circle as its vertex. An inscribed angle is formed by using two chords as its rays, and its vertex is a point on the circle itself. Finally, a **circumscribed angle** has a vertex that is a point outside the circle and rays that intersect with the circle. Some relationships exist between these types of angles, and, in order to define these relationships, arc measure must be understood. An **arc** of a circle is a portion of the circumference. Finding the **arc measure** is the same as finding the degree measure of the central angle that intersects the circle to form the arc. The measure of an inscribed angle is half the measure of its intercepted arc. It's also true that the measure of a circumscribed angle is equal to 180 degrees minus the measure of the central angle that forms the arc in the angle.

A **tangent line** is a line that touches a curve at a single point without going through it. A **compass** and a **straight edge** are the tools necessary to construct a tangent line from a point *P* outside the circle to the circle. A tangent line is constructed by drawing a line segment from the center of the circle *O* to the point *P*, and then finding its midpoint *M* by bisecting the line segment. By using *M* as the center, a compass is used to draw a circle through points *O* and *P*. *N* is defined as the intersection of the two circles. Finally, a line segment is drawn through *P* and *N*. This is the tangent line. Each point on a circle has only one tangent line, which is perpendicular to the radius at that point. A line similar to a tangent line is a **secant line**. Instead of intersecting the circle at one point, a secant line intersects the circle at two points. A **chord** is a smaller portion of a secant line.

Here's an example:

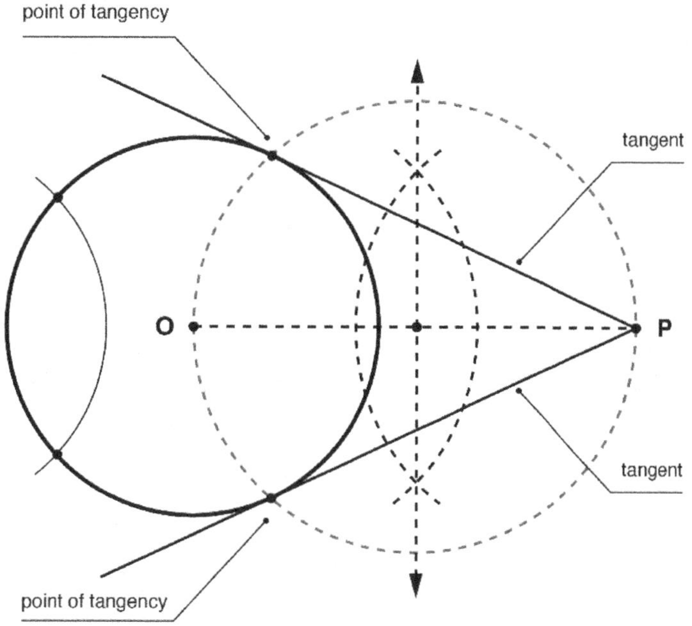

Geometric Relationships in the xy-Plane

Using Coordinate Geometry to Represent and Examine the Properties of Geometric Shapes

Many important formulas and equations exist in geometry that use coordinates. The distance between two points (x_1, y_1) and (x_2, y_2) is:

$$d = \sqrt{(x_2 - x_1)^2 + (y_2 - y_1)^2}.$$

The slope of the line containing the same two points is $m = \frac{y_2 - y_1}{x_2 - x_1}$. Also, the midpoint of the line segment with endpoints (x_1, y_1) and (x_2, y_2) is:

$$M = \left(\frac{x_1 + x_2}{2}, \frac{y_1 + y_2}{2}\right)$$

Two lines can be parallel, perpendicular, or neither. If two lines are **parallel**, they have the same slope. This is proven using the idea of similar triangles. Consider the following diagram with two parallel lines, L1 and L2:

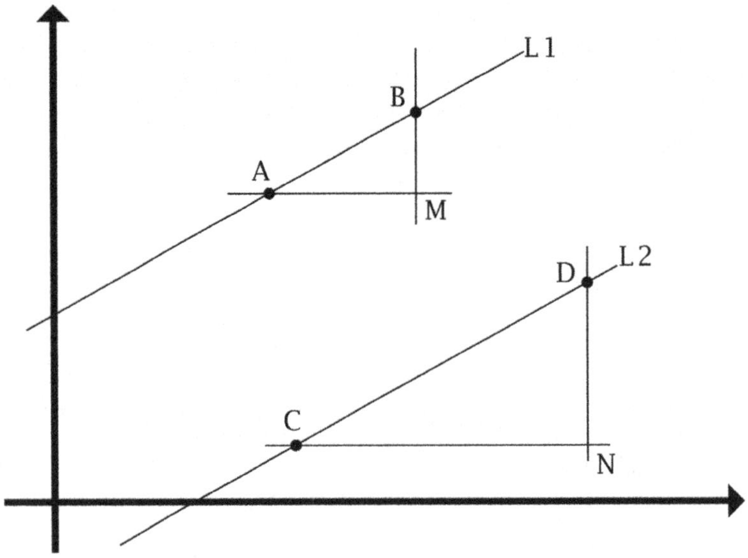

A and B are points on L1, and C and D are points on L2. Right triangles are formed with vertex M and N where lines BM and DN are parallel to the *y*-axis and AM and CN are parallel to the x-axis. Because all three sets of lines are parallel, the triangles are similar. Therefore, $\frac{\text{BM}}{\text{DN}} = \frac{\text{MA}}{\text{NC}}$. This shows that the rise/run is equal for lines L1 and L2. Hence, their slopes are equal.

Secondly, if two lines are **perpendicular**, the product of their slopes equals -1. This means that their slopes are negative reciprocals of each other. Consider two perpendicular lines, *l* and *n*:

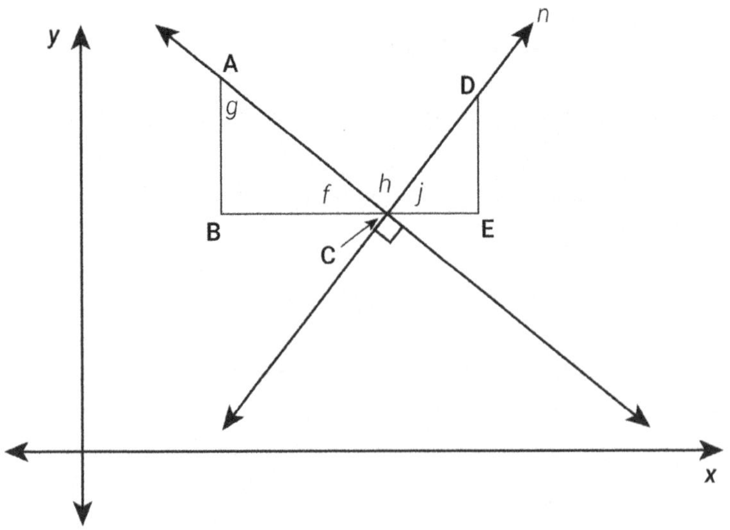

Right triangles ABC and CDE are formed so that lines BC and CE are parallel to the x-axis, and AB and DE are parallel to the *y*-axis. Because line BE is a straight line, angles $f + h + i = 180\ degrees$. However, angle *h* is a right angle, so $f + j = 90\ degrees$. By construction, $f + g = 90$, which means that $g = j$. Therefore, because angles $B = E$ and $g = j$, the triangles are similar and $\frac{\text{AB}}{\text{BC}} = \frac{\text{CE}}{\text{DE}}$. Because slope is equal to rise/run, the slope of line *l* is $-\frac{AB}{BC}$ and the slope of line *n* is $\frac{DE}{CE}$. Multiplying the slopes together gives:

$$-\frac{AB}{BC} \cdot \frac{DE}{CE} = -\frac{CE}{DE} \cdot \frac{DE}{CE} = -1$$

This proves that the product of the slopes of two perpendicular lines equals -1. Both parallel and perpendicular lines can be integral in many geometric proofs, so knowing and understanding their properties is crucial for problem-solving.

The Distance Between Two Points

The side lengths of a shape plotted on the coordinate plane or the length of a line segment can be found by plugging the endpoints into the distance formula between two ordered pairs (x_1, y_1) and (x_2, y_2).

As a reminder, this is the **distance formula**:

$$d = \sqrt{(x_2 - x_1)^2 + (y_2 - y_1)^2}$$

The distance formula is derived from the Pythagorean theorem. If working with a polygon, once the side lengths are found, they can be added together to obtain the perimeter of the given polygon. Simplifications can be made for specific shapes such as squares and equilateral triangles. For example, one side length can be multiplied by 4 to obtain the perimeter of a square. Also, one side length can be multiplied by 3 to obtain the perimeter of an equilateral triangle. A similar technique can be used to calculate areas. For polygons, both side length and height can be found by using the same distance formula. Areas of triangles and quadrilaterals are straightforward through the use of $A = \frac{1}{2}bh$ or $A = bh$, depending on the shape. To find the area of other polygons, their shapes can be partitioned into rectangles and triangles. The areas of these simpler shapes can be calculated and then added together to find the total area of the polygon.

The Midpoint of a Line Segment

If a line segment with endpoints (x_1, y_1) and (x_2, y_2) is partitioned into two equal parts, the formula for midpoint is used. Recall this formula is $M = \left(\frac{x_1+x_2}{2}, \frac{y_1+y_2}{2}\right)$ and the ratio of line segments is 1:1. For example, a line segment with points $(-2, 3)$ and $(4, -5)$ has a midpoint of $M = \left(\frac{-2+4}{2}, \frac{3+(-5)}{2}\right) = (1, -1)$. However, if the ratio needs to be anything other than 1:1, a different formula must be used. Consider a ratio that is $a{:}\,b$. This means the desired point that partitions the line segment is $\frac{a}{a+b}$ of the way from (x_1, y_1) to (x_2, y_2). The actual formula for the coordinate is $\left(\frac{bx_1+ax_2}{a+b}, \frac{by_1+ay_2}{a+b}\right)$.

Transformations

A **transformation** occurs when a shape is altered in the plane where it exists. There are three major types of transformation: translations, reflections, and rotations. A **translation** consists of shifting a shape in one direction. A **reflection** results when a shape is transformed over a line to its mirror image. Finally, a **rotation** occurs when a shape moves in a circular motion around a specified point. The object can be turned clockwise or counterclockwise and, if rotated 360 degrees, returns to its original location.

The three major types of transformations preserve distance and angle measurement. The shapes stay the same, but they are moved to another place in the plane. Therefore, the distance between any two points on the shape doesn't change. Also, any original angle measure between two line segments doesn't change. However, there are transformations that don't preserve distance and angle measurements, including those that don't preserve the original shape. For example, transformations that involve stretching and shrinking shapes don't preserve distance and angle measures. In these cases, the input variables are multiplied by either a number greater than one (**stretch**) or less than one (**shrink**).

A **rigid motion** is a transformation that preserves distance and length. Every line segment in the resulting image is congruent to the corresponding line segment in the pre-image. Congruence between two figures means a series of transformations (or a rigid motion) can be defined that maps one of the figures onto the other. Basically, two figures are congruent if they have the same shape and size.

Translation means that all points in the figure are moved in the same direction by the same distance. In other words, the figure is slid in some fixed direction. Of course, while the entire figure is slid by the same distance, this does not change any of the measurements of the figures involved. The result will have the same distances and angles as the original figure.

In terms of Cartesian coordinates, a translation means a shift of each of the original points (x, y) by a fixed amount in the *x* and *y* directions, to become $(x + a, y + b)$.

For reflection to occur, a line in the plane is specified, called the **line of reflection**. Then, take each point and flip it over the line so that it is the same distance from the line but on the opposite side of it. This does not change any of the distances or angles involved, but it does reverse the order in which everything appears.

To reflect something over the *x*-axis, the points (x, y) are sent to $(x, -y)$. To reflect something over the *y*-axis, the points (x, y) are sent to the points $(-x, y)$. Flipping over other lines is not something easy to express in Cartesian coordinates. However, by drawing the figure and the line of reflection, the distance to the line and the original points can be used to find the reflected figure.

Example: Reflect this triangle with vertices (-1, 0), (2, 1), and (2, 0) over the *y*-axis. The pre-image is shown below.

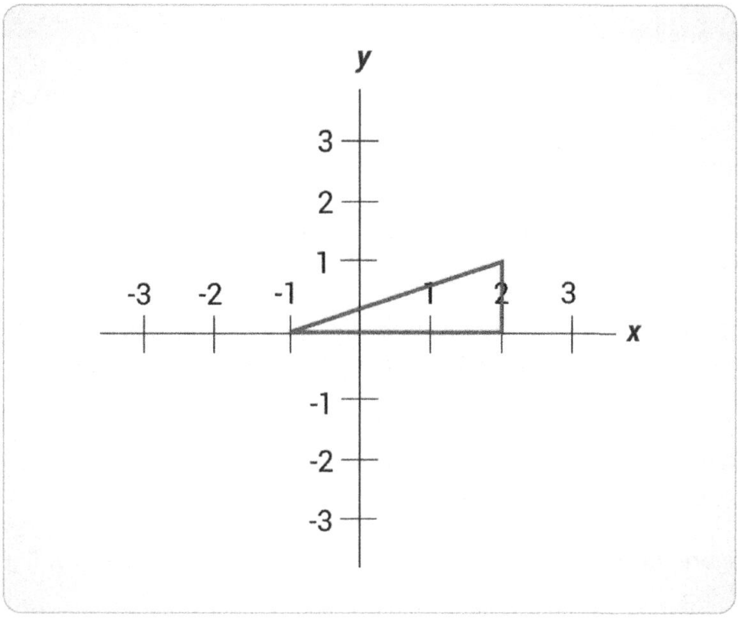

To do this, flip the *x* values of the points involved to the negatives of themselves, while keeping the *y* values the same. The image is shown here.

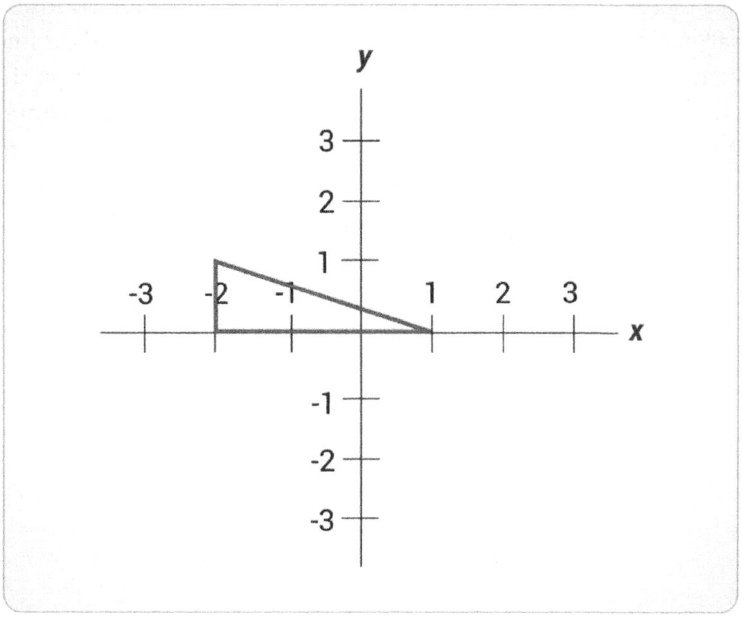

The new vertices will be (1, 0), (-2, 1), and (-2, 0).

For rotation, pick a center point, then rotate every vertex along a circle around that point by the same angle. This procedure is also not easy to express in Cartesian coordinates, and this is not a requirement on this test. However, as with reflections, it's helpful to draw the figures and see what the result of the rotation would look like. This transformation can be performed using a compass and protractor.

Each one of these transformations can be performed on the coordinate plane without changes to the original dimensions or angles.

If two figures in the plane involve the same distances and angles, they are called **congruent figures**. In other words, two figures are congruent when they go from one form to another through reflection, rotation, and translation, or a combination of these.

Remember that rotation and translation will give back a new figure that is identical to the original figure, but reflection will give back a mirror image of it.

To recognize that a figure has undergone a rotation, check to see that the figure has not been changed into a mirror image, but that its orientation has changed (that is, whether the parts of the figure now form different angles with the x and y axes).

To recognize that a figure has undergone a translation, check to see that the figure has not been changed into a mirror image, and that the orientation remains the same.

To recognize that a figure has undergone a reflection, check to see that the new figure is a mirror image of the old figure.

Keep in mind that sometimes a combination of translations, reflections, and rotations may be performed on a figure.

A shape is dilated, or a **dilation** occurs, when each side of the original image is multiplied by a given scale factor. If the scale factor is less than 1 and greater than 0, the dilation contracts the shape, and the resulting shape is smaller. If the scale factor equals 1, the resulting shape is the same size, and the dilation is a rigid motion. Finally, if the scale factor is greater than 1, the resulting shape is larger, and the dilation expands the shape. The **center of dilation** is the point where the distance from it to any point on the new shape equals the scale factor times the distance from the center to the corresponding point in the pre-image. Dilation isn't an isometric transformation because distance isn't preserved. However, angle measure, parallel lines, and points on a line all remain unchanged.

The following figure is an example of translation, rotation, dilation, and reflection:

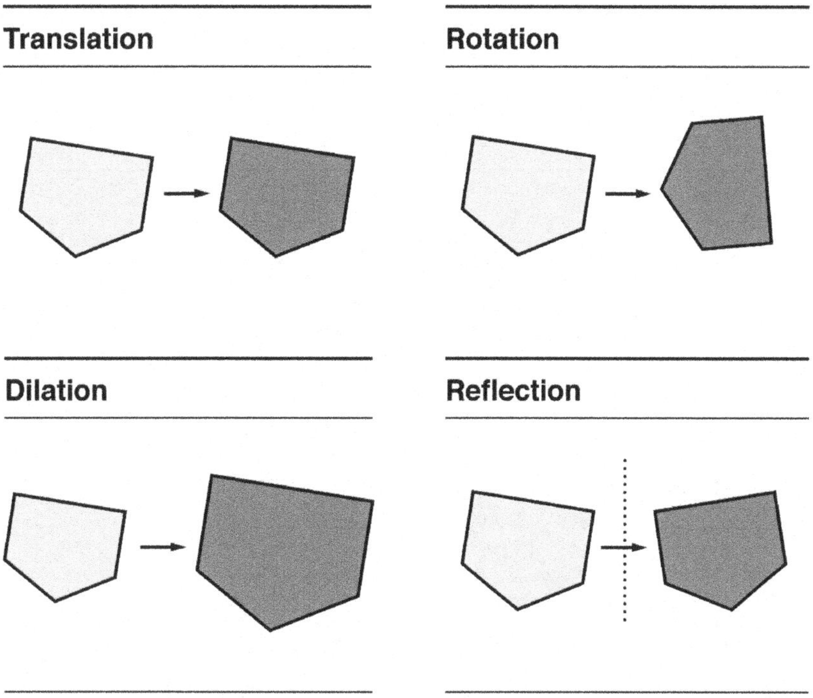

The easiest example is to dilate around the origin. In this case, multiply the *x* and *y* coordinates by a **scale factor**, k, sending points (x, y) to (kx, ky).

As an example, draw a dilation of the following triangle, whose vertices will be the points (-1, 0), (1, 0), and (1, 1).

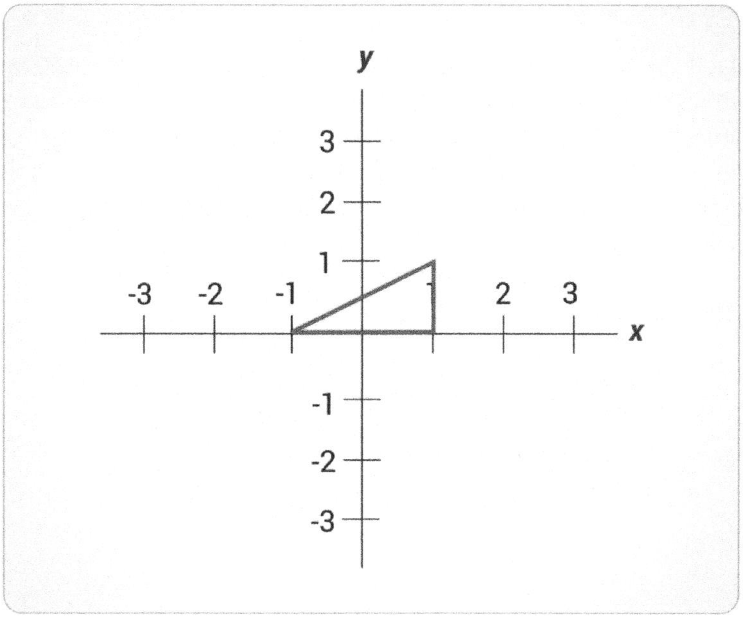

For this problem, dilate by a scale factor of 2, so the new vertices will be (-2, 0), (2, 0), and (2, 2).

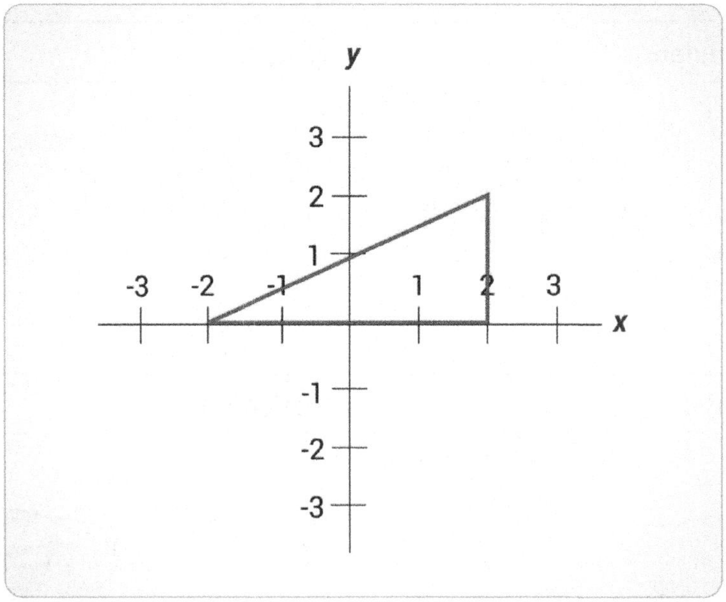

Note that after a dilation, the distances between the vertices of the figure will have changed, but the angles remain the same. The two figures that are obtained by dilation, along with possibly translation, rotation, and reflection, are all **similar** to one another. Another way to think of this is that similar figures have the same number of vertices and edges, and their angles are all the same. Similar figures have the same basic shape but are different in size.

Two figures are congruent if there is a rigid motion that can map one figure onto the other. Therefore, all pairs of sides and angles within the image and pre-image must be congruent. For example, in triangles, each pair of the three sides and three angles must be congruent. Similarly, in two four-sided figures, each pair of the four sides and four angles must be congruent.

Two figures are similar if there is a combination of translations, reflections, rotations, and dilations, which maps one figure onto the other. The difference between congruence and similarity is that dilation can be used in similarity. Therefore, side lengths between each shape can differ. However, angle measure must be preserved within this definition. If two polygons differ in size so that the lengths of corresponding line segments differ by the same factor, but corresponding angles have the same measurement, they are similar.

Systems of Measurement

The U.S. Customary and Metric Systems of Measurement

Measurement is how an object's length, width, height, weight, and so on, are quantified. Measurement is related to counting, but it is a more refined process.

The United States customary system and the metric system each consist of distinct units to measure lengths and volume of liquids. The U.S. customary units for length, from smallest to largest, are: inch (in), foot (ft), yard (yd), and mile (mi). The metric units for length, from smallest to largest, are: millimeter (mm), centimeter (cm), decimeter (dm), meter (m), and kilometer (km). The relative size of each unit of length is shown below.

U.S. Customary	Metric	Conversion
12in = 1ft	10mm = 1cm	1in = 2.54cm
36in = 3ft = 1yd	10cm = 1dm(decimeter)	1m ≈ 3.28ft ≈ 1.09yd
5,280ft = 1,760yd = 1mi	100cm = 10dm = 1m	1mi ≈ 1.6km
	1000m = 1km	

The U.S. customary units for volume of liquids, from smallest to largest, are: fluid ounces (fl oz), cup (c), pint (pt), quart (qt), and gallon (gal). The metric units for volume of liquids, from smallest to largest, are: milliliter (mL), centiliter (cL), deciliter (dL), liter (L), and kiloliter (kL). The relative size of each unit of liquid volume is shown below.

U.S. Customary	Metric	Conversion
8fl oz = 1c	10mL = 1cL	1pt ≈ 0.473L
2c = 1pt	10cL = 1dL	1L ≈ 1.057qt
4c = 2pt = 1qt	1,000mL = 100cL = 10dL = 1L	1gal ≈ 3.785L
4qt = 1gal	1,000L = 1kL	

The U.S. customary system measures weight (how strongly Earth is pulling on an object) in the following units, from least to greatest: ounce (oz), pound (lb), and ton. The metric system measures mass (the quantity of matter within an object) in the following units, from least to greatest: milligram (mg), centigram (cg), gram (g), kilogram (kg), and metric ton (MT).

The relative sizes of each unit of weight and mass are shown below.

U.S. Measures of Weight	Metric Measures of Mass
16oz = 1lb	10mg = 1cg
2,000lb = 1 ton	100cg = 1g
	1,000g = 1kg
	1,000kg = 1MT

Note that weight and mass DO NOT measure the same thing.

Time is measured in the following units, from shortest to longest: second (sec), minute (min), hour (h), day (d), week (wk), month (mo), year (yr), decade, century, millennium. The relative sizes of each unit of time is shown below.

- 60sec = 1min
- 60min = 1h
- 24hr = 1d
- 7d = 1wk
- 52wk = 1yr

- 12mo = 1yr
- 10yr = 1 decade
- 100yrs = 1 century
- 1,000yrs = 1 millennium

Some units of measure are represented as square or cubic units depending on the solution. For example, perimeter is measured in linear units, area is measured in square units, and volume is measured in cubic units.

Also be sure to use the most appropriate unit for the thing being measured. A building's height might be measured in feet or meters while the length of a nail might be measured in inches or centimeters. Additionally, for SI units, the prefix should be chosen to provide the most succinct available value. For example, the mass of a bag of fruit would likely be measured in kilograms rather than grams or milligrams, and the length of a bacteria cell would likely be measured in micrometers rather than centimeters or kilometers.

Problems that involve measurements of length, time, volume, etc. are generally dependent upon understanding how to manipulate between various units of measurement, as well as understanding their equivalencies.

Identifying and utilizing the proper units for the scenario requires knowing how to apply the conversion rates for money, length, volume, and mass. For example, given a scenario that requires subtracting 8 inches from $2\frac{1}{2}$ feet, both values should first be expressed in the same unit (they could be expressed $\frac{2}{3}$ft & $2\frac{1}{2}$ft, or 8in and 30in). The desired unit for the answer may also require converting back to another unit.

Consider the following scenario: A parking area along the river is only wide enough to fit one row of cars and is $\frac{1}{2}$ kilometers long. The average space needed per car is 5 meters. How many cars can be parked along the river? First, all measurements should be converted to similar units: $\frac{1}{2}$km = 500m. The operation(s) needed should be identified. Because the problem asks for the number of cars, the total space should be divided by the space per car. 500 meters divided by 5 meters per car yields a total of 100 cars. Written as an expression, the meters unit cancels and the cars unit is left: $\frac{500m}{5m/car}$ the same as $500m \times \frac{1\ car}{5m}$ yields 100 cars.

For an example manipulating time, Maria is scheduled to take a 90-minute test for her English class. It takes her 25 minutes to get ready and 40 minutes to ride the bus to school. If she begins to get ready at 1:10 p.m., what time will she be finished taking the test?

To find the ending time, all of the elapsed minutes must be totaled and then converted to hours.

$$25 + 40 + 90 = 155 \text{ minutes}$$

The conversion necessary for this problem is that 1 hour = 60 minutes.

The total number of minutes must be converted into hours and minutes, by dividing the total number of minutes by 60.

$$155 \div 60 = 2\ R\ 35$$

The remainder is stated as minutes. So, the total elapsed time is 2 hours and 35 minutes. If Maria begins to get ready at 1:10 p.m., 2 hours from that time is 3:10 p.m., and an additional 35 minutes would add up to 3:45 p.m. Maria can expect to be finished with everything 2 hours and 35 minutes later, at 3:45 p.m.

When measuring length, choosing the right tool to perform the measurement requires determining whether United States customary units or metric units are desired, and having a grasp of the approximate length of each unit and the approximate length of each tool. The measurement can still be performed by trial and error without the knowledge of the approximate size of the tool.

For example, to determine the length of a room in feet, a United States customary unit, various tools can be used for this task. These include a ruler (typically 12 inches/1 foot long), a yardstick (3 feet/1 yard long), or a tape measure displaying feet (typically either 25 feet or 50 feet). Because the length of a room is much larger than the length of a ruler or a yardstick, a tape measure should be used to perform the measurement.

Converting Units of Measurement

Converting measurements in different units between the two systems can be difficult because they follow different rules. The best method is to look up an English to Metric system conversion factor and then use a series of equivalent fractions to set up an equation to convert the units of one of the measurements into those of the other. The table below lists some common conversion values that are useful for problems involving measurements with units in both systems:

English System	**Metric System**
1 inch	2.54 cm
1 foot	0.3048 m
1 yard	0.914 m
1 mile	1.609 km
1 ounce	28.35 g
1 pound	0.454 kg
1 fluid ounce	29.574 mL
1 quart	0.946 L
1 gallon	3.785 L

Consider the example where a scientist wants to convert 6.8 inches to centimeters. One method for converting units is to write and solve a proportion. The arrangement of values in a proportion is extremely important. The table above is used to find that there are 2.54 centimeters in every inch, so the following equation should be set up and solved:

$$\frac{6.8\ in}{1} \times \frac{2.54\ cm}{1\ in} = 17.272\ cm$$

Notice how the inches in the numerator of the initial figure and the denominator of the conversion factor cancel out. (This equation could have been written simply as $6.8\ in\ \times\ 2.54\ cm = 17.272\ cm$, but it was shown in detail to illustrate the steps). The goal in any conversion equation is to set up the fractions so that the units you are trying to convert from cancel out and the units you desire remain.

For a more complicated example, consider converting 2.15 kilograms into ounces. The first step is to convert kilograms into grams and then grams into ounces. Note that the measurement you begin with does not have to be put in a fraction.

So in this case, 2.15 kg is by itself although it's technically the numerator of a fraction:

$$2.15\ kg \times \frac{1000g}{kg} = 2150\ g$$

Then, use the conversion factor from the table to convert grams to ounces:

$$2150g \times \frac{1\ oz}{28.35g} = 75.8\ oz$$

Now suppose that a problem requires converting 20 fluid ounces to cups. To do so, a proportion can be written using the conversion rate of 8fl oz = 1c with *x* representing the missing value. The proportion can be written in any of the following ways:

$$\frac{1}{8} = \frac{x}{20}\left(\frac{c\ for\ conversion}{fl\ oz\ for\ conversion} = \frac{unknown\ c}{fl\ oz\ given}\right); \frac{8}{1} = \frac{20}{x}\left(\frac{fl\ oz\ for\ conversion}{c\ for\ conversion} = \frac{fl\ oz\ given}{unknown\ c}\right);$$

$$\frac{1}{x} = \frac{8}{20}\left(\frac{c\ for\ conversion}{unknown\ c} = \frac{fl\ oz\ for\ conversion}{fl\ oz\ given}\right); \frac{x}{1} = \frac{20}{8}\left(\frac{unknown\ c}{c\ for\ conversion} = \frac{fl\ oz\ given}{fl\ oz\ for\ conversion}\right)$$

To solve the proportion, the ratios are cross-multiplied and the resulting equation is solved. When cross-multiplying, all four proportions above will produce the same equation: $(8)(x) = (20)(1) \rightarrow 8x = 20$. Divide by 8 to isolate the variable *x*, the result is *x* = 2.5. The variable *x* represented the unknown number of cups. Therefore, the conclusion is that 20 fluid ounces converts (is equal) to 2.5 cups.

Sometimes converting units requires writing and solving more than one proportion. Suppose an exam question asks to determine how many hours are in 2 weeks. Without knowing the conversion rate between hours and weeks, this can be determined knowing the conversion rates between weeks and days, and between days and hours. First, weeks are converted to days, then days are converted to hours. To convert from weeks to days, the following proportion can be written:

$$\frac{7}{1} = \frac{x}{2}\left(\frac{days\ conversion}{weeks\ conversion} = \frac{days\ unknown}{weeks\ given}\right)$$

Cross-multiplying produces: $(7)(2) = (x)(1) \rightarrow 14 = x$. Therefore, 2 weeks is equal to 14 days. Next, a proportion is written to convert 14 days to hours:

$$\frac{24}{1} = \frac{x}{14}\left(\frac{conversion\ hours}{conversion\ days} = \frac{unknown\ hours}{given\ days}\right)$$

Cross-multiplying produces: $(24)(14) = (x)(1) \rightarrow 336 = x$. Therefore, the answer is that there are 336 hours in 2 weeks.

Geometric Constructions

Formal Geometric Constructions

The tools needed to make formal geometric constructions are a compass, a ruler, paper folding, or geometry software. These tools can be used to copy or bisect (split into two equal halves) a line segment or angle. They can also replicate a specific shape and construct perpendicular lines, a perpendicular bisector (a line that bisects) of a line segment, and a line parallel to a given line through a specified point.

Beginning with formal geometric constructions, various geometric figures and shapes can be built. Definitions and theorems for lines and angles can be used in parallel with geometric constructions to build shapes such as equilateral triangles, squares, and rectangular hexagons inscribed in a circle. Definitions of shapes involving congruence of sides and angles within each type of figure must be understood and used in parallel with constructing congruent line segments, parallel and perpendicular lines, and congruent angles.

Diagonals are lines (excluding sides) that connect two vertices within a polygon. **Mutually bisecting diagonals** intersect at their midpoints. Parallelograms, rectangles, squares, and rhombuses have mutually bisecting diagonals. However, trapezoids don't have such lines. **Perpendicular diagonals** occur when they form four right triangles at their point of intersection. Squares and rhombuses have perpendicular diagonals, but trapezoids, rectangles, and parallelograms do not. Finally, **perpendicular bisecting** diagonals (also known as **perpendicular bisectors**) form four right triangles at their point of intersection, but this intersection is also the midpoint of the two lines. Both rhombuses and squares have perpendicular bisecting angles, but trapezoids, rectangles, and parallelograms do not. Knowing these definitions can help tremendously in problems that involve both angles and diagonals.

A **pentagon** is a five-sided figure. A six-sided shape is a *hexagon.* A seven-sided figure is classified as a **heptagon,** and an eight-sided figure is called an *octagon.* An important characteristic is whether a polygon is regular or irregular. If it's *regular*, the side lengths and angle measurements are all equal. An **irregular** polygon has unequal side lengths and angle measurements. Mathematical problems involving polygons with more than four sides usually involve side length and angle measurements. The sum of all internal angles in a polygon equals $180(n-2)$ degrees, where n is the number of sides. Therefore, the total of all internal angles in a pentagon is 540 degrees because there are five sides so $180(5-2) = 540$ degrees. Unfortunately, area formulas don't exist for polygons with more than four sides. However, their shapes can be split up into triangles, and the formula for area of a triangle can be applied and totaled to obtain the area for the entire figure.

Probability, Statistics, and Discrete Mathematics

Interpreting and Analyzing Data Presented in Various Forms

Interpreting Various Displays of Data

A set of data can be visually displayed in various forms allowing for quick identification of characteristics of the set. **Histograms**, such as the one shown below, display the number of data points (vertical axis) that fall into given intervals (horizontal axis) across the range of the set. The histogram below displays the heights of black cherry trees in a certain city park. Each rectangle represents the number of trees with heights between a given five-point span. For example, the furthest bar to the right indicates that two trees are between 85 and 90 feet. Histograms can describe the center, spread, shape, and any unusual characteristics of a data set.

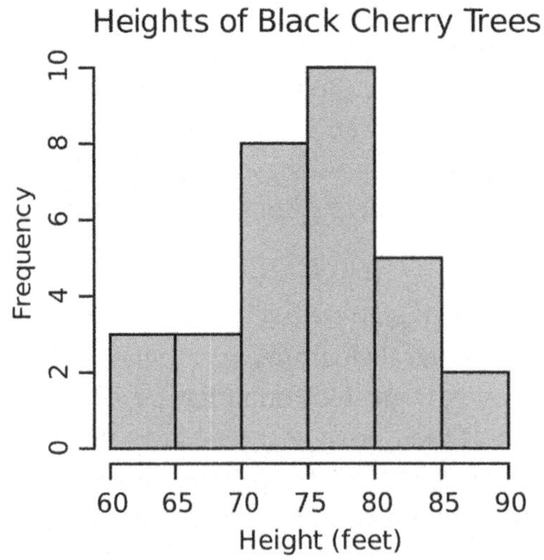

A **box plot**, also called a **box-and-whisker plot**, divides the data points into four groups and displays the five number summary for the set, as well as any outliers. The five number summary consists of:

- The lower extreme: the lowest value that is not an outlier
- The higher extreme: the highest value that is not an outlier
- The median of the set: also referred to as the second quartile or Q_2
- The first quartile or Q_1: the median of values below Q_2
- The third quartile or Q_3: the median of values above Q_2

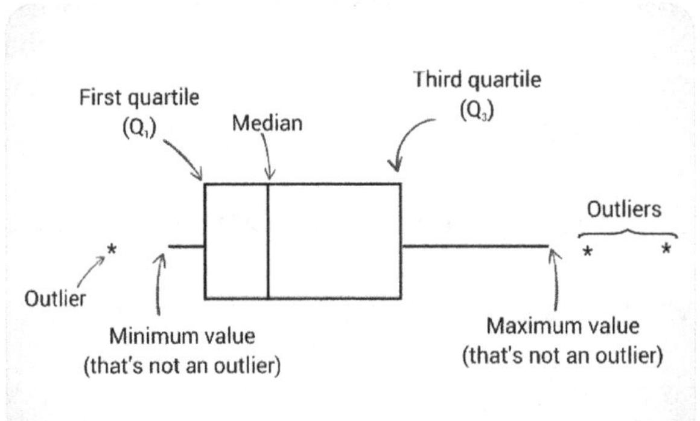

Suppose the box plot displays IQ scores for 12th grade students at a given school. The five number summary of the data consists of: lower extreme (67); upper extreme (127); Q_2or median (100); Q_1 (91); Q_3 (108); and outliers (135 and 140). Although all data points are not known from the plot, the points are divided into four quartiles each, including 25% of the data points. Therefore, 25% of students scored between 67 and 91, 25% scored between 91 and 100, 25% scored between 100 and 108, and 25% scored between 108 and 127. These percentages include the normal values for the set and exclude the outliers. This information is useful when comparing a given score with the rest of the scores in the set.

A **scatter plot** is a mathematical diagram that visually displays the relationship or connection between two variables. The independent variable is placed on the *x*-axis, or horizontal axis, and the dependent variable is placed on the *y*-axis, or vertical axis. When visually examining the points on the graph, if the points model a linear relationship, or a **line of best-fit** can be drawn through the points with the points relatively close on either side, then a correlation exists. If the line of best-fit has a positive slope (rises from left to right), then the variables have a positive correlation. If the line of best-fit has a negative slope (falls from left to right), then the variables have a negative correlation. If a line of best-fit cannot be drawn, then no correlation exists. A positive or negative correlation can be categorized as strong or weak, depending on how closely the points are graphed around the line of best-fit.

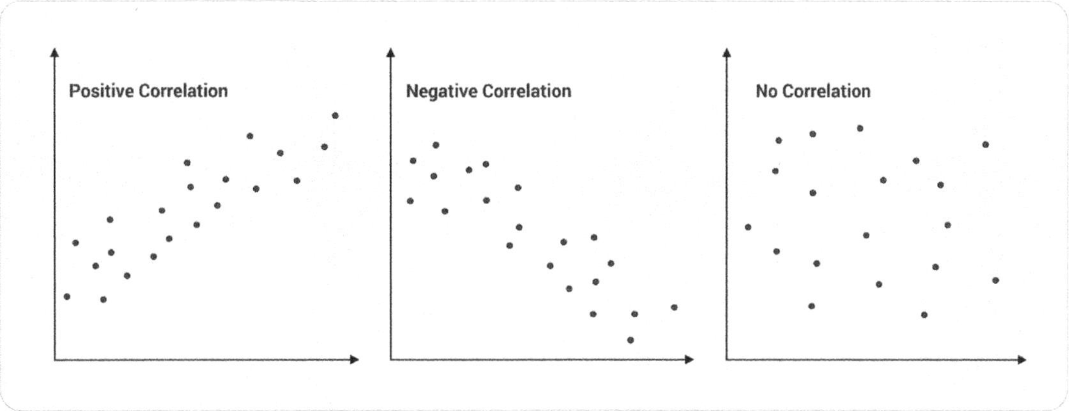

Drawing Conclusions Based on Graphical Displays

In an experiment, variables are the key to analyzing data, especially when data is in a graph or table. Variables can represent anything, including objects, conditions, events, and amounts of time.

Covariance is a general term referring to how two variables move in relation to each other. Take for example an employee that gets paid by the hour. For them, hours worked and total pay have a positive covariance. As hours worked increases, so does pay.

Constant variables remain unchanged by the scientist across all trials. Because they are held constant for all groups in an experiment, they aren't being measured in the experiment, and they are usually ignored. Constants can either be controlled by the scientist directly like the nutrition, water, and sunlight given to plants, or they can be selected by the scientist specifically for an experiment like using a certain animal species or choosing to investigate only people of a certain age group.

Independent variables are also controlled by the scientist, but they are the same only for each group or trial in the experiment. Each group might be composed of students that all have the same color of car or each trial may be run on different soda brands. The independent variable of an experiment is what is being indirectly tested because it causes change in the dependent variables.

Dependent variables experience change caused by the independent variable and are what is being measured or observed. For example, college acceptance rates could be a dependent variable of an experiment that sorted a large sample of high school students by an independent variable such as test scores. In this experiment, the scientist groups the high school students by the independent variable (test scores) to see how it affects the dependent variable (their college acceptance rates).

Note that most variables can be held constant in one experiment, but also serve as the independent variable or a dependent variable in another. For example, when testing how well a fertilizer aids plant growth, its amount of sunlight should be held constant for each group of plants, but if the experiment is being done to determine the proper amount of sunlight a plant should have, the amount of sunlight is an independent variable because it is necessarily changed for each group of plants.

An **X-Y diagram**, also known as a **scatter diagram**, visually displays the relationship between two variables. The independent variable is placed on the *x-axis*, or horizontal axis, and the dependent variable is placed on the *y-axis*, or vertical axis.

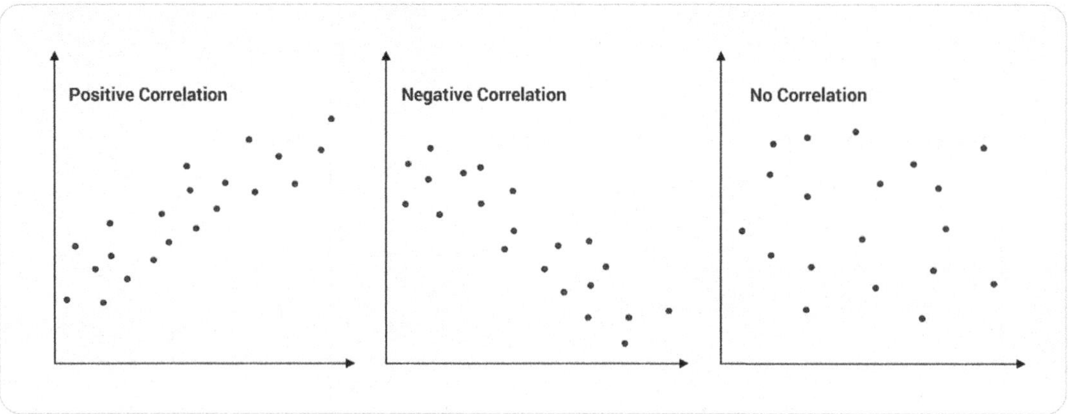

As shown in the figures above, an X-Y diagram may result in positive, negative, or no correlation between the two variables. In the first scatter plot, as the Y factor increases, the X factor increases as well. The opposite is true as well: as the X factor increases the Y factor also increases. Thus there is a positive correlation because one factor appears to positively affect the other factor

It's important to note, however, that a positive correlation between two variables doesn't equate to a cause-and-effect relationship. For example, a positive correlation between labor hours and units produced may not equate to a cause and effect relationship between the two. Any instance of correlation only indicates how likely the presence of one variable is in the instance of another. The variables should be further analyzed to determine which, if any, other variables (i.e. quality of employee work) may contribute to the positive correlation.

Comparing data sets within statistics can mean many things. The first way to compare data sets is by looking at the center and spread of each set. The **center of a data** set can mean two things: median or mean. The **median** is the value that's halfway into each data set, and it splits the data into two intervals. The **mean** is the average value of the data within a set. It's calculated by adding up all of the data in the set and dividing the total by the number of data points. Outliers can significantly impact the mean. Additionally, two completely different data sets can have the same mean. For example, a data set with values ranging from 0 to 100 and a data set with values ranging from 44 to 56 can both have means of 50. The first data set has a much wider range, which is known as the **spread** of the data. This measures how varied the data is within each set.

Interpreting Competing Data

Be careful of questions with competing studies. These questions will ask the student to interpret which of two studies shows the greater amount or the higher rate of change between two results.

Here's an example. A research facility runs studies on two different antibiotics: Drug A and Drug B. The Drug A study includes 1,000 participants and cures 600 people. The Drug B study includes 200 participants and cures 150 people. Which drug is more successful?

The first step is to determine the percentage of each drug's rate of success. Drug A was successful in curing 60% of participants, while Drug B achieved a 75% success rate. Thus, Drug B is more successful based on these studies, even though it cured fewer people.

Sample size and experiment consistency should also be considered when answering questions based on competing studies. Is one study significantly larger than the other? In the antibiotics example, the Drug A study is five times larger than Drug B. Thus, Drug B's higher efficacy (desired result) could be a result of the smaller sample size, rather than the quality of drug.

Consistency between studies is directly related to sample size. Let's say the research facility elects to conduct more studies on Drug B. In the next study, there are 400 participants, and 200 are cured. The success rate of the second study is 50%. The results are clearly inconsistent with the first study, which means more testing is needed to determine the drug's efficacy. A hallmark of mathematical or scientific research is repeatability. Studies should be consistent and repeatable, with an appropriately large sample size, before drawing extensive conclusions.

Representing Data in Various Forms

Constructing Graphs

Various graphs can be used to visually represent a given set of data. Each type of graph requires a different method of arranging data points and different calculations of the data. To construct a histogram, the range of the data points is divided into equal intervals. The frequency for each interval is then determined, which reveals how many points fall into each interval. A graph is constructed with the vertical axis representing the frequency and the horizontal axis representing the intervals. The lower value of each interval should be labeled along the horizontal axis. Finally, for each interval, a bar is drawn from the lower value of each interval to the lower value of the next interval with a height equal to the frequency of the interval. Because of the intervals, histograms do not have any gaps between bars along the horizontal axis.

To construct a box (or box-and-whisker) plot, the five number summary for the data set is calculated as follows: the **second quartile (Q_2)** is the median of the set. The **first quartile (Q_1)** is the median of the values below Q_2. The **third quartile (Q_3)** is the median of the values above Q_2. The **upper extreme** is the highest value in the data set if it is not an **outlier** (greater than 1.5 times the interquartile range Q_3 - Q_1). The lower extreme is the least value in the data set if it is not an outlier (more than 1.5 times lower than the interquartile range). To construct the box-and-whisker plot, each value is plotted on a number line, along with any outliers. The **box** consists of Q_1 and Q_3 as its *top* and *bottom* and Q_2 as the dividing line inside the box. The **whiskers** extend from the lower extreme to Q_1 and from Q_3 to the upper extreme.

Box Plot

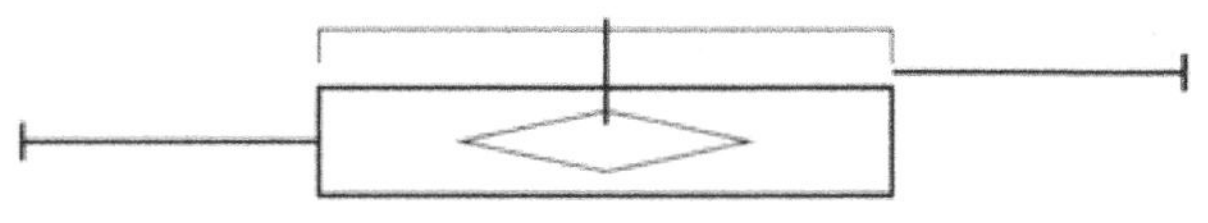

A scatter plot displays the relationship between two variables. Values for the independent variable, typically denoted by *x*, are paired with values for the dependent variable, typically denoted by *y*. Each set of corresponding values are written as an ordered pair (*x*, *y*). To construct the graph, a coordinate grid is labeled with the *x*-axis representing the independent variable and the *y*-axis representing the dependent variable. Each ordered pair is graphed.

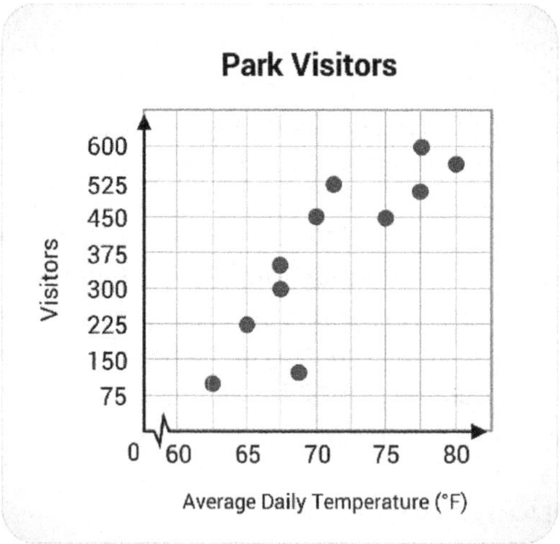

Like a scatter plot, a **line graph** compares variables that change continuously, typically over time. Paired data values (ordered pair) are plotted on a coordinate grid with the *x*- and *y*-axis representing the variables. A line is drawn from each point to the next, going from left to right. The line graph below displays cell phone use for given years (two variables) for men, women, and both sexes (three data sets).

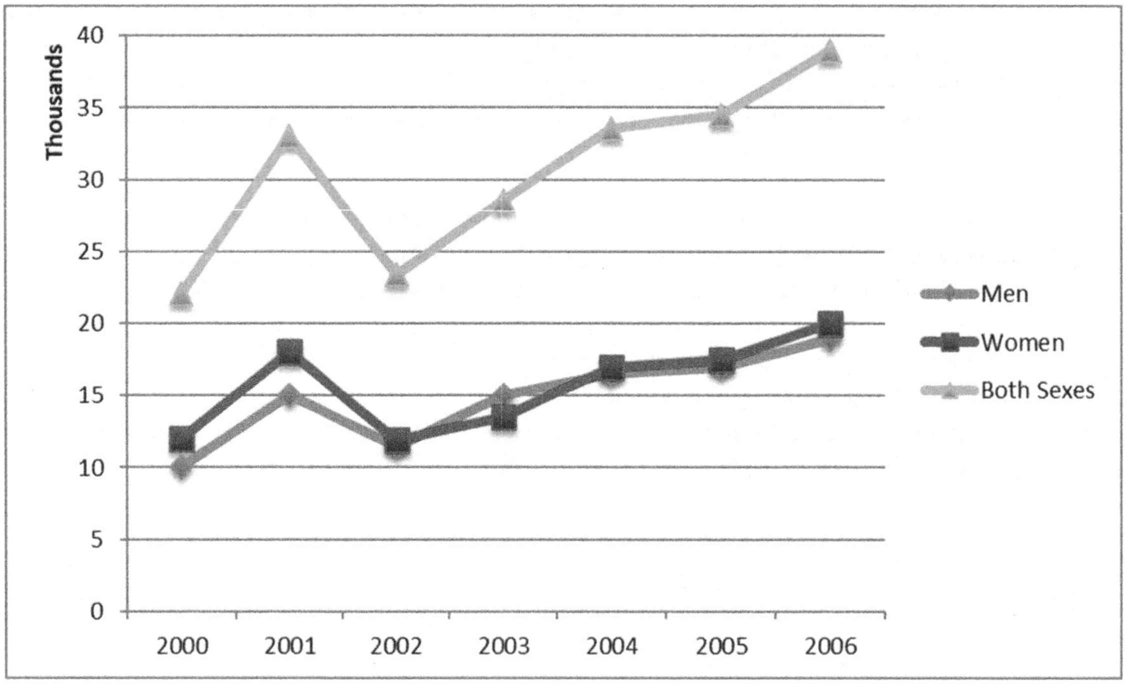

A **line plot**, also called **dot plot**, displays the frequency of data (numerical values) on a number line. To construct a line plot, a number line is used that includes all unique data values. It is marked with x's or dots above the value the number of times that the value occurs in the data set.

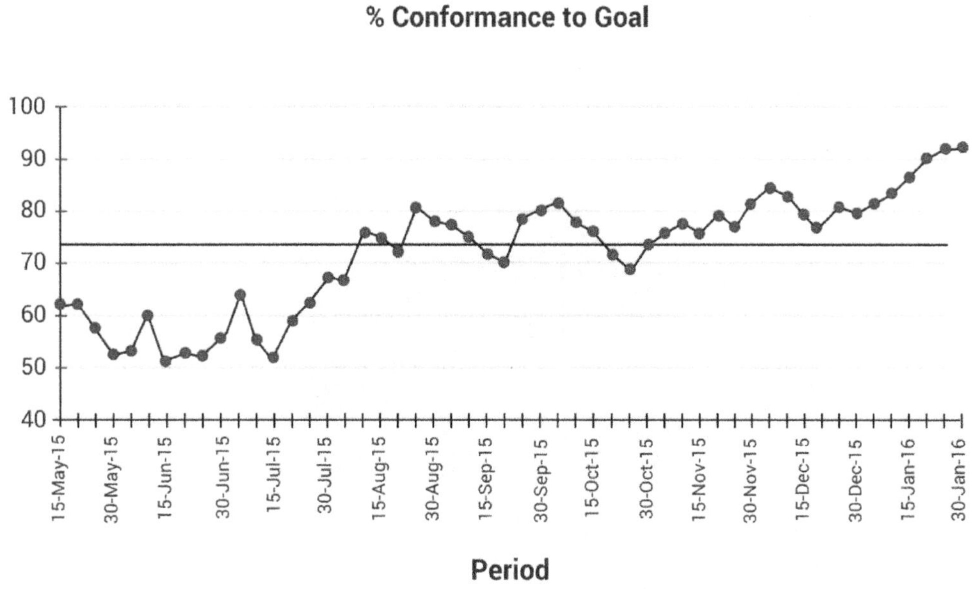

A **bar graph** is a diagram in which the quantity of items within a specific classification is represented by the height of a rectangle. Each type of classification is represented by a rectangle of equal width. Here is an example of a bar graph:

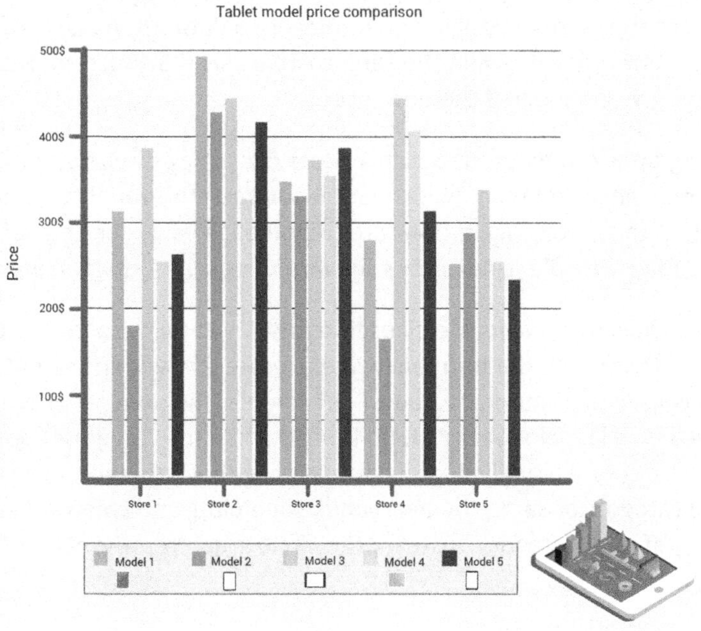

A **circle graph**, also called a **pie chart**, shows categorical data with each category representing a percentage of the whole data set. To make a circle graph, the percent of the data set for each category must be determined. To do so, the frequency of the category is divided by the total number of data points and converted to a percent. For example, if 80 people were asked what their favorite sport is and 20 responded basketball, basketball makes up 25% of the data ($\frac{20}{80} = 0.25 = 25\%$). Each category in a data set is represented by a slice of the circle proportionate to its percentage of the whole.

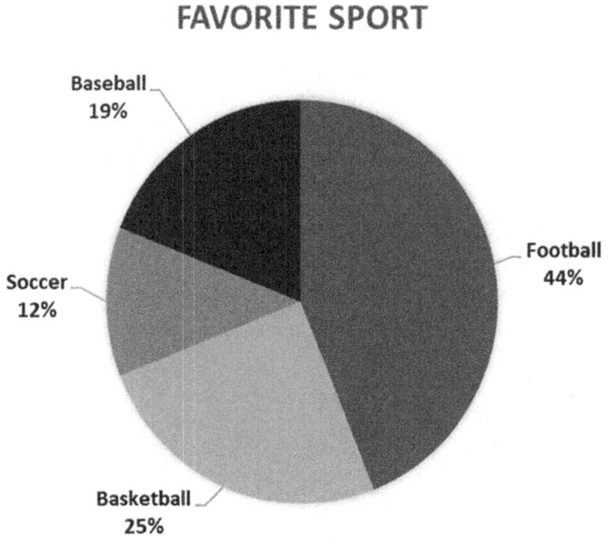

Choosing an Appropriate Graph Based on Data

Choosing the appropriate graph to display a data set depends on what type of data is included in the set and what information must be displayed. Histograms and box plots can be used for data sets consisting of individual values across a wide range. Examples include test scores and incomes. Histograms and box plots will indicate the center, spread, range, and outliers of a data set. A histogram will show the shape of the data set, while a box plot will divide the set into quartiles (25% increments), allowing for comparison between a given value and the entire set.

Scatter plots and line graphs can be used to display data consisting of two variables. Examples include height and weight, or distance and time. A correlation between the variables is determined by examining the points on the graph. Line graphs are used if each value for one variable pairs with a distinct value for the other variable. Line graphs show relationships between variables.

Line plots, bar graphs, and circle graphs are all used to display categorical data, such as surveys. Line plots and bar graphs both indicate the frequency of each category within the data set. A line plot is used when the categories consist of numerical values. For example, the number of hours of TV watched by individuals is displayed on a line plot. A bar graph is used when the categories consists of words. For example, the favorite ice cream of individuals is displayed with a bar graph. A circle graph can be used to display either type of categorical data. However, unlike line plots and bar graphs, a circle graph does not indicate the frequency of each category. Instead, the circle graph represents each category as its percentage of the whole data set.

Probability Models

Counting Techniques

There are many counting techniques that can help solve problems involving counting possibilities. For example, the **Addition Principle** states that if there are m choices from Group 1 and n choices from Group 2, then $n + m$ is the total number of choices possible from Groups 1 and 2. For this to be true, the groups can't have any choices in common. The **Multiplication Principle** states that if Process 1 can be completed n ways and Process 2 can be completed m ways, the total number of ways to complete both Process 1 and Process 2 is $n \times m$. For this rule to be used, both processes must be independent of each other. Counting techniques also involve permutations. A **permutation** is an arrangement of elements in a set for which order must be considered. For example, if three letters from the alphabet are chosen, ABC and BAC are two different permutations. The multiplication rule can be used to determine the total number of possibilities. If each letter can't be selected twice, the total number of possibilities is $26 \times 25 \times 24 = 15{,}600$. A formula can also be used to calculate this total. In general, the notation $P(n, r)$ represents the number of ways to arrange r objects from a set of n and, the formula is $P(n, r) = \frac{n!}{(n-r)!}$. In the previous example, $P(26, 3) = \frac{26!}{23!} = 15{,}600$. Contrasting permutations, a **combination** is an arrangement of elements in which order doesn't matter. In this case, ABC and BAC are the same combination. In the previous scenario, there are six permutations that represent each single combination. Therefore, the total number of possible combinations is $15{,}600 \div 6 = 2{,}600$. In general, $C(n, r)$ represents the total number of combinations of n items selected r at a time where order doesn't matter, and the formula is $C(n, r) = \frac{n!}{(n-r)!\,r!}$. Therefore, the following relationship exists between permutations and combinations: $C(n, r) = \frac{P(n,r)}{n!} = \frac{P(n,r)}{P(r,r)}$.

Probability for an event is equal to the number of outcomes in that event divided by the total number of outcomes in the sample space. For example, consider rolling a 6-sided die. The probability of rolling an odd number is $\frac{3}{6}$, or $\frac{1}{2}$, because there are 3 odd numbers on the die.

The **fundamental counting principle** states that if there are m possible ways for an event to occur, and n possible ways for a second event to occur, there are $m \times n$ possible ways for both events to occur. For example, there are two events that can occur after flipping a coin and six events that can occur after rolling a die, so there are $2 \times 6 = 12$ total possible event scenarios if both are done simultaneously. This principle can be used to find probabilities involving finite sample spaces and independent trials because it calculates the total number of possible outcomes. For this principle to work, the events must be independent of each other.

Independent and Dependent Events

Probability is a measure of how likely an event is to occur. Probability is written as a fraction between zero and one. If an event has a probability of zero, the event will never occur. If an event has a probability of one, the event will definitely occur. If the probability of an event is closer to zero, the event is unlikely to occur. If the probability of an event is closer to one, the event is more likely to occur. For example, a probability of $\frac{1}{2}$ means that the event is equally as likely to occur as it is not to occur. An example of this is tossing a coin. To calculate the probability of an event, the number of favorable outcomes is divided by the number of total outcomes. For example, suppose you have 2 raffle tickets out of 20 total tickets sold. The probability that you win the raffle is calculated: $\frac{number\ of\ favorable\ outcomes}{total\ numberof\ outcomes} = \frac{2}{20} = \frac{1}{10}$ (always reduce fractions). Therefore, the probability of winning the raffle is $\frac{1}{10}$ or 0.1.

Chance is the measure of how likely an event is to occur, written as a percent. If an event will never occur, the event has a 0% chance. If an event will certainly occur, the event has a 100% chance. If an event will sometimes occur, the event has a chance somewhere between 0% and 100%. To calculate chance, probability is calculated and the fraction is converted to a percent.

The probability of multiple events occurring can be determined by multiplying the probability of each event. For example, suppose you flip a coin with heads and tails, and roll a six-sided dice numbered one through six. To find the probability that you will flip heads AND roll a two, the probability of each event is determined and those fractions are multiplied. The probability of flipping heads is $\frac{1}{2}\left(\frac{1\ side\ with\ heads}{2\ sides\ total}\right)$ and the probability of rolling a two is $\frac{1}{6}\left(\frac{1\ side\ with\ a\ 2}{6\ total\ sides}\right)$. The probability of flipping heads AND rolling a 2 is: $\frac{1}{2} \times \frac{1}{6} = \frac{1}{12}$.

The above scenario with flipping a coin and rolling a dice is an example of independent events. Independent events are circumstances in which the outcome of one event does not affect the outcome of the other event. Conversely, dependent events are ones in which the outcome of one event affects the outcome of the second event. Consider the following scenario: a bag contains 5 black marbles and 5 white marbles. What is the probability of picking 2 black marbles without replacing the marble after the first pick?

The probability of picking a black marble on the first pick is $\frac{5}{10}\left(\frac{5\ black\ marbles}{10\ total\ marbles}\right)$. Assuming that a black marble was picked, there are now 4 black marbles and 5 white marbles for the second pick. Therefore, the probability of picking a black marble on the second pick is $\frac{4}{9}\left(\frac{4\ black\ marbles}{9\ total\ marbles}\right)$. To find the probability of picking two black marbles, the probability of each is multiplied: $\frac{5}{10} \times \frac{4}{9} = \frac{20}{90} = \frac{2}{9}$.

Geometric Probability

Probability deals with the likelihood that a certain event will occur. Geometric probability is a specific type of probability; it deals with the likelihood of getting a point in a particular region of a geometric entity (such as a length, area, or volume) out of the total length, area, or volume of the entity. While the formula to calculate probability generally involves the ratio of the number of favorable events to the number of total events, the formula for calculating geometric probability involves the ratio of desired area to the total area (here, "area" is used loosely to mean a geometric quantity, including length, volume, and true area).

Imagine the following figure is an unusually-shaped dart board:

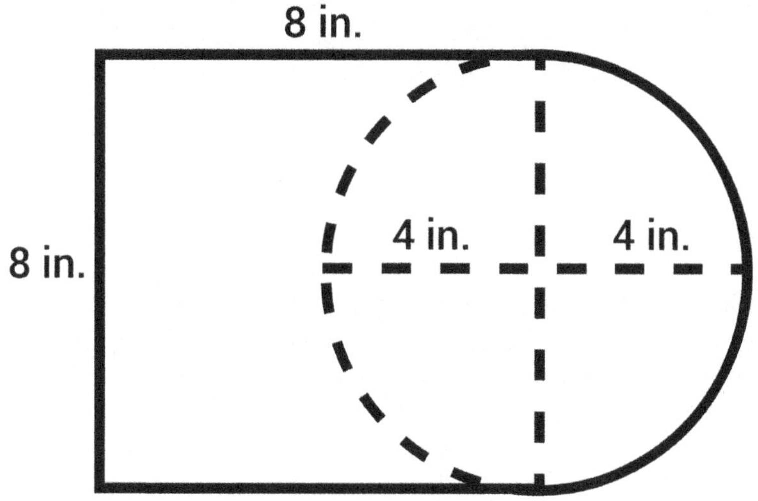

A potential geometric probability question could entail calculating the likelihood that a dart thrown at this unique dart board lands in the bottom right quadrant of the circle. The formula to calculate a geometric probability is:

$$P = \frac{area\ of\ desired\ region}{total\ area}$$

Therefore, the first step is to calculate the total area of the figure (to get the denominator) and then the area of ¼ of the circle (the lower right quadrant), which is the numerator.

This figure is formed from a circle and a square. The diameter of the circle is the length of one of the sides of the square. Because half of the circle overlaps the square, only one half of the circle adds to the footprint of the area of the figure. Essentially, because of the overlap, the area of this figure is the area of the square plus the area of half of the circle. The area of a square is calculated by squaring the length of the side, so this square has an area of 8 inches × 8 inches, or 64 square inches. The area of a circle is calculated from the formula $A = \pi r^2$. Here, 4 inches is the radius, so the area is 16π. We only need the area of half the circle, so that is 8π. Using the approximation of 3.14 for π, $A = 8 \times 3.14$ or 25.12 square inches. Adding this to the area of the circle yields 89.12 square inches. This is the total area. The area of one quarter of the circle is the area we just found (25.12 square inches), which was the area of half the circle, divided by 2 since there are two quarters in every half. Dividing 25.12 square inches by 2 yields 12.56 square inches. Thus, the area of the lower right quarter of the circle is 12.56 square inches. This is the numerator for our geometric probability formula. Plugging in both values will now yield the probability:

$$P = \frac{area\ of\ desired\ region}{total\ area} = \frac{12.56\ in^2}{89.12\ in^2} = 0.141$$

Measures of Central Tendency

Solving for the Mean and Weighted Average of a Given Set of Data

The **average** or **mean** of a collection of numbers is given by adding those numbers together and then dividing by the total number of values. A **weighted average** or **weighted mean** is given by adding the numbers multiplied by their weights, then dividing by the sum of the weights:

$$\frac{w_1x_1 + w_2x_2 + w_3x_3 \dots + w_nx_n}{w_1 + w_2 + w_3 + \dots + w_n}$$

An **ordinary average** is a weighted average where all the weights are 1.

Determining and Interpreting Mean, Median, and Mode

The **center** of a set of data (statistical values) can be represented by its mean, median, or mode. These are sometimes referred to as **measures of central tendency**. The **mean** is the average of the data set. The mean can be calculated by adding the data values and dividing by the sample size (the number of data points). Suppose a student has test scores of 93, 84, 88, 72, 91, and 77. To find the mean, or average, the scores are added and the sum is divided by 6 because there are 6 test scores:

$$\frac{93 + 84 + 88 + 72 + 91 + 77}{6} = \frac{505}{6} = 84.17$$

Given the mean of a data set and the sum of the data points, the sample size can be determined by dividing the sum by the mean. Suppose you are told that Kate averaged 12 points per game and scored a total of 156 points for the season. The number of games that she played (the sample size or the number of data points) can be determined by dividing the total points (sum of data points) by her average (mean of data points): $\frac{156}{12} = 13$. Therefore, Kate played in 13 games this season.

If given the mean of a data set and the sample size, the sum of the data points can be determined by multiplying the mean and sample size. Suppose you are told that Tom worked 6 days last week for an average of 5.5 hours per day. The total number of hours worked for the week (sum of data points) can be determined by multiplying his daily average (mean of data points) by the number of days worked (sample size): $5.5 \times 6 = 33$. Therefore, Tom worked a total of 33 hours last week.

The **median** of a data set is the value of the data point in the middle when the sample is arranged in numerical order. To find the median of a data set, the values are written in order from least to greatest. The lowest and highest values are simultaneously eliminated, repeating until the value in the middle remains. Suppose the salaries of math teachers are: $35,000; $38,500; $41,000; $42,000; $42,000; $44,500; $49,000. The values are listed from least to greatest to find the median. The lowest and highest values are eliminated until only the middle value remains. Repeating this step three times reveals a median salary of $42,000. If the sample set has an even number of data points, two values will remain after all others are eliminated. In this case, the mean of the two middle values is the median. Consider the following data set: 7, 9, 10, 13, 14, 14. Eliminating the lowest and highest values twice leaves two values, 10 and 13, in the middle. The mean of these values $\left(\frac{10+13}{2}\right)$ is the median. Therefore, the set has a median of 11.5.

The **mode** of a data set is the value that appears most often. A data set may have a single mode, multiple modes, or no mode. If different values repeat equally as often, multiple modes exist. If no value repeats, no mode exists. Consider the following data sets:

- A: 7, 9, 10, 13, 14, 14
- B: 37, 44, 33, 37, 49, 44, 51, 34, 37, 33, 44
- C: 173, 154, 151, 168, 155

Set A has a mode of 14. Set B has modes of 37 and 44. Set C has no mode.

The range of a data set is the difference between the highest and the lowest values in the set. The range can be considered the span of the data set. To determine the range, the smallest value in the set is subtracted from the largest value. The ranges for the data sets A, B, and C above are calculated as follows: A: $14 - 7 = 7$; B: $51 - 33 = 18$; C: $173 - 151 = 22$.

Changing all values of a data set in a consistent way produces predictable changes in the measures of the center and range of the set. A linear transformation changes the original value into the new value by either adding a given number to each value, multiplying each value by a given number, or both. Adding (or subtracting) a given value to each data point will increase (or decrease) the mean, median, and any modes by the same value. However, the range will remain the same due to the way that range is calculated. Multiplying (or dividing) a given value by each data point will increase (or decrease) the mean, median, and any modes, and the range by the same factor.

Consider the following data set, call it set *P*, representing the price of different cases of soda at a grocery store: \$4.25, \$4.40, \$4.75, \$4.95, \$4.95, \$5.15. The mean of set *P* is \$4.74. The median is \$4.85. The mode of the set is \$4.95. The range is \$0.90. Suppose the state passes a new tax of \$0.25 on every case of soda sold. The new data set, set *T*, is calculated by adding \$0.25 to each data point from set *P*. Therefore, set *T* consists of the following values: \$4.50, \$4.65, \$5.00, \$5.20, \$5.20, \$5.40. The mean of set *T* is \$4.99. The median is \$5.10. The mode of the set is \$5.20. The range is \$.90. The mean, median and mode of set *T* is equal to \$0.25 added to the mean, median, and mode of set *P*. The range stays the same.

Now suppose, due to inflation, the store raises the cost of every item by 10 percent. Raising costs by 10 percent is calculated by multiplying each value by 1.1. The new data set, set *I*, is calculated by multiplying each data point from set *T* by 1.1. Therefore, set *I* consists of the following values: \$4.95, \$5.12, \$5.50, \$5.72, \$5.72, \$5.94. The mean of set *I* is \$5.49. The median is \$5.61. The mode of the set is \$5.72. The range is \$0.99. The mean, median, mode, and range of set *I* is equal to 1.1 multiplied by the mean, median, mode, and range of set *T* because each increased by a factor of 10 percent.

Common Features of a Data Set

A set of data can be described in terms of its center, spread, shape and any unusual features. The **center** of a data set can be measured by its mean, median, or mode. The **spread** of a data set refers to how far the data points are from the center (mean or median). The spread can be measured by the range or the quartiles and interquartile range. A data set with all its data points clustered around the center will have a small spread. A data set covering a wide range of values will have a large spread.

When a data set is displayed as a histogram or frequency distribution plot, the shape indicates if a sample is normally distributed, symmetrical, or has measures of skewness or kurtosis. When graphed, a data set with a normal distribution will resemble a bell curve.

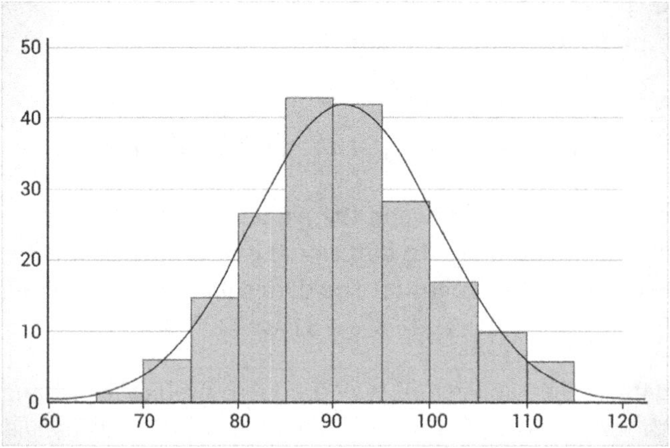

If the data set is **symmetrical**, each half of the graph when divided at the center is a mirror image of the other. If the graph has fewer data points to the right, the data is **skewed right**. If it has fewer data points to the left, the data is **skewed left**.

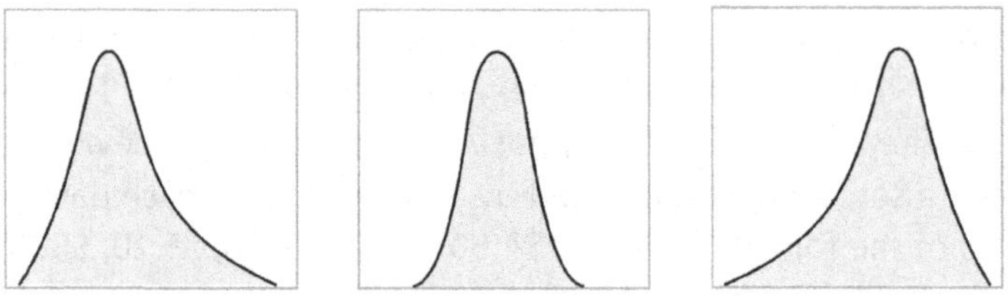

Kurtosis is a measure of whether the data is heavy-tailed with a high number of outliers, or light-tailed with a low number of outliers.

A description of a data set should include any unusual features such as gaps or outliers. A **gap** is a span within the range of the data set containing no data points. An **outlier** is a data point with a value either extremely large or extremely small when compared to the other values in the set.

Choosing an Appropriate Measure of Central Tendency

Measures of central tendency, namely mean, median, and mode, describe characteristics of a set of data. Specifically, they are intended to represent a *typical* value in the set by identifying a central position of the set. Depending on the characteristics of a specific set of data, different measures of central tendency are more indicative of a typical value in the set.

When a data set is grouped closely together with a relatively small range and the data is spread out somewhat evenly, the mean is an effective indicator of a typical value in the set. Consider the following data set representing the height of sixth grade boys in inches: 61 inches, 54 inches, 58 inches, 63 inches, 58 inches. The mean of the set is 58.8 inches. The data set is grouped closely (the range is only 9 inches) and the values are spread relatively evenly (three values below the mean and two values above the mean). Therefore, the mean value of 58.8 inches is an effective measure of central tendency in this case.

When a data set contains a small number of values either extremely large or extremely small when compared to the other values, the mean is not an effective measure of central tendency. Consider the following data set representing annual incomes of homeowners on a given street: $71,000; $74,000; $75,000; $77,000; $340,000. The mean of this set is $127,400. This figure does not indicate a typical value in the set, which contains four out of five values between $71,000 and $77,000. The median is a much more effective measure of central tendency for data sets such as these. Finding the middle value diminishes the influence of outliers, or numbers that may appear out of place, like the $340,000 annual income. The median for this set is $75,000 which is much more typical of a value in the set.

The mode of a data set is a useful measure of central tendency for categorical data when each piece of data is an option from a category. Consider a survey of 31 commuters asking how they get to work with results summarized below.

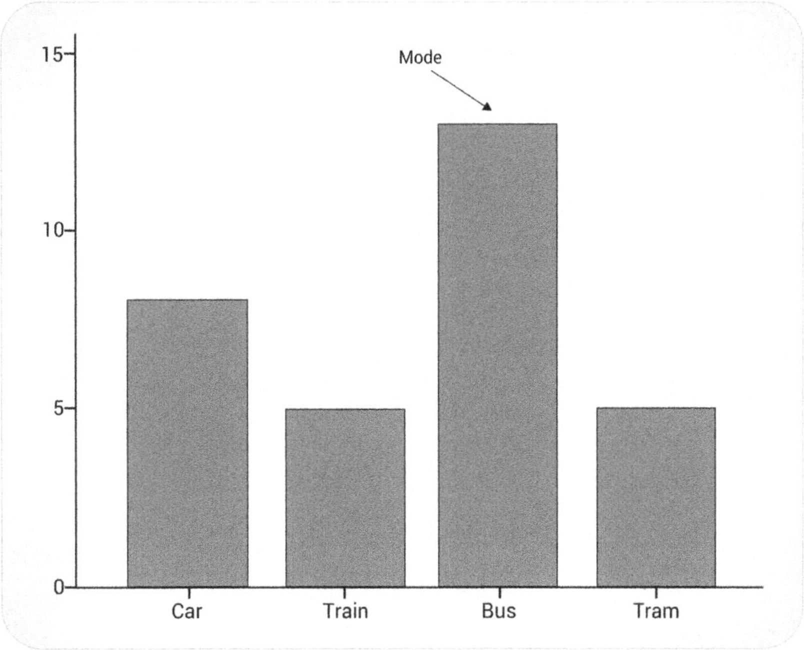

The mode for this set represents the value, or option, of the data that repeats most often. This indicates that the bus is the most popular method of transportation for the commuters.

Modeling and Solving Problems Using Simple Diagrams, Flowcharts, or Algorithms

Constructing, Using, and Interpreting Simple Diagrams to Solve Problems

In solving multi-step problems, the first step is to line up the available information. Then, try to decide what information the problem is asking to be found. Sketching out a diagram to illustrate what is known and what is unknown can be helpful in tackling the problem and determining what mathematical equations or processes need to be used to find the solution.

For example, you can construct a strip diagram to display the known information along with any information to be calculated. Finally, the missing information can be represented by a **variable** (a letter from the alphabet that represents a number) in a mathematical equation that the student can solve.

For example, Delilah collects stickers and her friends gave her some stickers to add to her current collection. Joe gave her 45 stickers, and Aimee gave her 2 times the number of stickers that Joe gave Delilah. How many stickers did Delilah have to start with, if after her friends gave her more stickers, she had a total of 187 stickers?

In order to solve this, the given information must first be sorted out. Joe gives Delilah 45 stickers, Aimee gives Delilah 2 times the number Joe gives (2 × 45), and the end total of stickers is 187.

A strip diagram represents these numbers as follows:

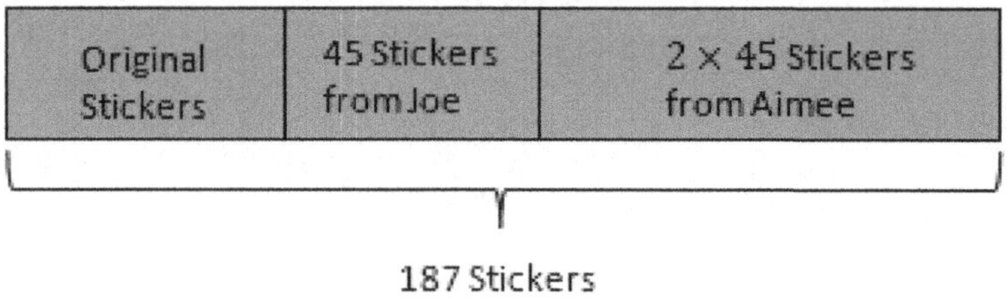

The entire situation can be modeled by this equation, using the variable *s* to stand for the original number of stickers:

$$s + 45 + (2 \times 45) = 187.$$

Solving for *s* would give the solution, as follows:

$$s + 45 + 90 = 187$$

$$s + 135 = 187$$

$$s + 135 - 135 = 187 - 135$$

$$s = 52 \text{ stickers.}$$

Applying a Given Algorithm to Solve a Problem

When attempting to solve a math problem, it's important to apply the correct algorithm. It is much more difficult to determine what algorithm is necessary when solving word problems, because the necessary operations and equations are typically not provided. In these instances, the test taker must translate the words in the problem into true mathematical statements that can be solved.

Word problems can appear daunting, but don't let the wording intimidate you. No matter the scenario or specifics, the key to answering them is to translate the words into a math problem. Always keep in mind what the question is asking and what operations could lead to that answer. The following word problem resembles one of the question types most frequently encountered on the exam.

Walter's Coffee Shop sells a variety of drinks and breakfast treats.

Price List	
Hot Coffee	$2.00
Slow-Drip Iced Coffee	$3.00
Latte	$4.00
Muffin	$2.00
Crepe	$4.00
Egg Sandwich	$5.00

Costs	
Hot Coffee	$0.25
Slow-Drip Iced Coffee	$0.75
Latte	$1.00
Muffin	$1.00
Crepe	$2.00
Egg Sandwich	$3.00

Walter's utilities, rent, and labor costs him $500 per day. Today, Walter sold 200 hot coffees, 100 slow-drip iced coffees, 50 lattes, 75 muffins, 45 crepes, and 60 egg sandwiches. What was Walter's total profit today?

To accurately answer this type of question, determine the total cost of making his drinks and treats, then determine how much revenue he earned from selling those products. After arriving at these two totals, the profit is measured by deducting the total cost from the total revenue.

Walter's costs for today:

Item	Quantity	Cost Per Unit	Total Cost
Hot Coffee	200	$0.25	$50
Slow-Drip Iced Coffee	100	$0.75	$75
Latte	50	$1.00	$50
Muffin	75	$1.00	$75
Crepe	45	$2.00	$90
Egg Sandwich	60	$3.00	$180
Utilities, rent, and labor			$500
Total Costs			$1,020

Walter's revenue for today:

Item	Quantity	Revenue Per Unit	Total Revenue
Hot Coffee	200	$2.00	$400
Slow-Drip Iced Coffee	100	$3.00	$300
Latte	50	$4.00	$200
Muffin	75	$2.00	$150
Crepe	45	$4.00	$180
Egg Sandwich	60	$5.00	$300
Total Revenue			$1,530

Walter's Profit = *Revenue* – *Costs* = $1,530 – $1,020 = $510

This strategy is applicable to other question types. For example, calculating salary after deductions, balancing a checkbook, and calculating a dinner bill are common word problems similar to business planning. Just remember to use the correct operations. When a balance is increased, use addition.

When a balance is decreased, use subtraction. Common sense and organization are your greatest assets when answering word problems.

In general, when solving word problems (also called story problems), it's important to understand what is being asked and to properly set up the initial equation. Always read the entire problem through, and then separate out what information is given in the statement. Decide what you are being asked to find and label each quantity with a variable or constant. Then write an equation to determine the unknown variable. Remember to label answers; sometimes knowing what the answers' units can help eliminate other possible solutions.

When trying to solve any word problem, look for a series of key words indicating addition, subtraction, multiplication, or division to help you determine how to set up the problem:

Addition: add, altogether, together, plus, increased by, more than, in all, sum, and total

Subtraction: minus, less than, difference, decreased by, fewer than, remain, and take away

Multiplication: *times*, *twice*, *of*, *double*, and *triple*

Division: divided by, cut up, half, quotient of, split, and shared equally

If a question asks to give words to a mathematical expression and says "equals," then an = sign must be included in the answer. Similarly, "less than or equal to" is expressed by the inequality symbol ≤, and "greater than or equal" to is expressed as ≥. Furthermore, "less than" is represented by <, and "greater than" is expressed by >.

Example:

A store is having a spring sale, where everything is 70% off. You have $45.00 to spend. A jacket is regularly priced at $80.00. Do you have enough to buy the jacket and a pair of gloves, regularly priced at $20.00?

There are two ways to approach this.

Method 1:

Set up the equations to find the sale prices: the original price minus the amount discounted.
$80.00 - ($80.00 (0.70)) = sale cost of the jacket.
$20.00 – ($20.00 (0.70)) = sale cost of the gloves.
Solve for the sale cost.
$24.00 = sale cost of the jacket.
$6.00 = sale cost of the gloves.
Determine if you have enough money for both.
$24.00 + $6.00 = total sale cost.
$30.00 is less than $45.00, so you can afford to purchase both.

Method 2:

Determine the percent of the original price that you will pay.
100% – 70% = 30%

Set up the equations to find the sale prices.
$80.00 (0.30) = cost of the jacket.
$20.00 (0.30) = cost of the gloves.
Solve.
$24.00 = cost of the jacket.
$6.00 = cost of the gloves.
Determine if you have enough money for both.
$24.00 + $6.00 = total sale cost.
$30.00 is less than $45.00, so you can afford to purchase both.

Example:

Mary and Dottie team up to mow neighborhood lawns. If Mary mows 2 lawns per hour and the two of them can mow 17.5 lawns in 5 hours, how many lawns does Dottie mow per hour?

Given rate for Mary.

$$Mary = \frac{2\ lawns}{1\ hour}$$

Unknown rate of D for Dottie.

$$Dottie = \frac{D\ lawns}{1\ hour}$$

Given rate for both.

$$Total\ mowed\ together = \frac{17.5\ lawns}{5\ hours}$$

Set up the equation for what is being asked.

$$Mary + Dottie = total\ together.$$

Fill in the givens.

$$2 + D = \frac{17.5}{5}$$

Divide.

$$2 + D = 3.5$$

Subtract 2 from both sides to isolate the variable.

$$2 - 2 + D = 3.5 - 2$$

Solve and label Dottie's mowing rate.

$$D = 1.5\ lawns\ per\ hour$$

Practice Questions

1. A rectangle has a length that is 5 feet longer than three times its width. If the perimeter is 90 feet, what is the length in feet?

   a. 10
   b. 20
   c. 25
   d. 35

2. Ten students take a test. Five students get a 50. Four students get a 70. If the average score is 55, what was the last student's score?

   a. 20
   b. 40
   c. 50
   d. 60

3. An equilateral triangle has a perimeter of 18 feet. If a square whose sides have the same length as one side of the triangle is built, what will be the area of the square? Enter your answer in the box below.

4. The table below displays the number of three-year-olds at Kids First Daycare who are potty-trained and those who still wear diapers.

	Potty-trained	**Wear diapers**	
Boys	26	22	48
Girls	34	18	52
	60	40	

What is the probability that a three-year-old girl chosen at random from the school is potty-trained?

   a. 52 percent
   b. 34 percent
   c. 65 percent
   d. 57 percent

5. What is the value of x? Enter your answer in the box below.

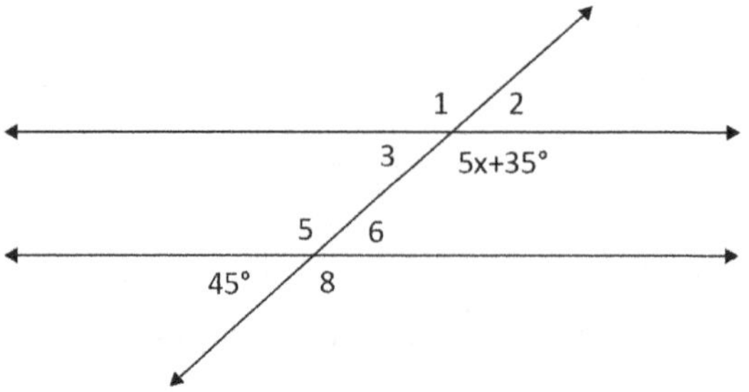

Answer Explanations

1. D: Denote the width as w and the length as l. Then, $l = 3w + 5$. The perimeter is $2w + 2l = 90$. Substituting the first expression for l into the second equation yields:

$$2(3w + 5) + 2w = 90$$

$$6w + 10 + 2w = 90$$

$$8w = 80$$

$$w = 10$$

Putting this into the first equation, it yields:

$$l = 3(10) + 5 = 35$$

2. A: Let the unknown score be x. The average will be:

$$\frac{5 \times 50 + 4 \times 70 + x}{10} = \frac{530 + x}{10} = 55$$

Multiply both sides by 10 to get $530 + x = 550$, or $x = 20$.

3. 36 square feet. An equilateral triangle has three sides of equal length, so if the total perimeter is 18 feet, each side must be 6 feet long. A square with sides of 6 feet will have an area of $6^2 = 36$ square feet.

4. C: The conditional frequency of a girl being potty-trained is calculated by dividing the number of potty-trained girls by the total number of girls: $34 \div 52 = 0.65$. To determine the conditional probability, multiply the conditional frequency by 100: $0.65 \times 100 = 65\%$.

5. 20. Because these are 2 parallel lines cut by a transversal, the angle with a measure of 45 degrees is equal to the measure of angle 6. Angle 6 and the angle labeled $5x + 35$ are supplementary to one another. The sum of these angles should be 180, so the following equation can be generated:

$$5x + 35 + 45 = 180$$

Solving for x, the sum of 35 and 45 is 80, which is then subtracted from 180 to yield 100. Dividing 100 by 5 gives the value of x, which is 20.

Practice Test

Arithmetic and Algebra

1. What is the value of the expression: $7^2 - 3 \times (4 + 2) + 15 \div 5$?
   a. 12.2
   b. 40.2
   c. 34
   d. 58.2

2. Solve the following:

$$(\sqrt{36} \times \sqrt{16}) - 3^2$$

3. In 2015, it was estimated that there were 7,350,000,000 people living on Earth. Express this value in scientific notation.
   a. 7.35×10^7
   b. 7.35×10^9
   c. 73.5×10^8
   d. 73.5×10^9

4. Express the solution to the following problem in decimal form:

$$\frac{3}{5} \times \frac{7}{10} \div \frac{1}{2}$$

   a. 0.042
   b. 84%
   c. 0.84
   d. 0.42

5. If Danny takes 48 minutes to walk 3 miles, how long should it take him to walk 5 miles maintaining the same speed?

6. The variable y is directly proportional to x. If $y = 3$ when $x = 5$, then what is y when $x = 20$?

a. 10
b. 12
c. 14
d. 16

7. Which of the following is a counterexample that disproves the statement, "All prime numbers are also odd numbers"?

a. 1
b. 2
c. 100
d. There is no counterexample because it is a true statement.

8. You measure the width of your door to be 36 inches. The true width of the door is 35.75 inches. What is the relative error in your measurement?

a. 0.7%
b. 0.007%
c. 0.99%
d. 0.1%

9. What is the value of x in the following equation?

$$13 - \frac{3x}{4} = -11$$

10. Kimberley earns $10 an hour babysitting, and after 10 p.m., she earns $12 an hour, with the amount paid being rounded to the nearest hour accordingly. On her last job, she worked from 5:30 p.m. to 11 p.m. In total, how much did Kimberley earn on her last job?

a. $45
b. $57
c. $62
d. $42

11. Solve this equation:

$$9x + x - 7 = 16 + 2x$$

a. $x = -4$

b. $x = 3$

c. $x = \frac{9}{8}$

d. $x = \frac{23}{8}$

12. Keith's bakery had 252 customers go through its doors last week. This week, that number increased to 378. Express this increase as a percentage.
   a. 26%
   b. 50%
   c. 35%
   d. 12%

13. If $4x - 3 = 5$, what is the value of x?

14. How will the following algebraic expression be simplified: $(5x^2 - 3x + 4) - (2x^2 - 7)$?
   a. x^5
   b. $3x^2 - 3x + 11$
   c. $3x^2 - 3x - 3$
   d. $x - 3$

15. Which of the following is the result of simplifying the expression: $\frac{4a^{-1}b^3}{a^4b^{-2}} \times \frac{3a}{b}$?
   a. $12a^3b^5$
   b. $12\frac{b^4}{a^4}$
   c. $\frac{12}{a^4}$
   d. $7\frac{b^4}{a}$

16. 40 gallons of milk was poured into two containers of different sizes. Which of the following represents the amount of milk poured into the smaller container in terms of the amount m poured into the larger container?
   a. $40 + m$
   b. $m - 40$
   c. $40 - m$
   d. $40/m$

17. Karen gets paid a weekly salary and a commission for every sale that she makes. The table below shows the number of sales and her pay for different weeks.

Sales	2	7	4	8
Pay	$380	$580	$460	$620

Which of the following equations represents Karen's weekly pay?
   a. $y = 90x + 200$
   b. $y = 90x - 200$
   c. $y = 40x + 300$
   d. $y = 40x - 300$

18. Which graph will be a line parallel to the graph of $y = 3x - 2$?
   a. $2y - 6x = 2$
   b. $y - 4x = 4$
   c. $3y = x - 2$
   d. $2x - 2y = 2$

19. Which inequality represents the values displayed on the number line?

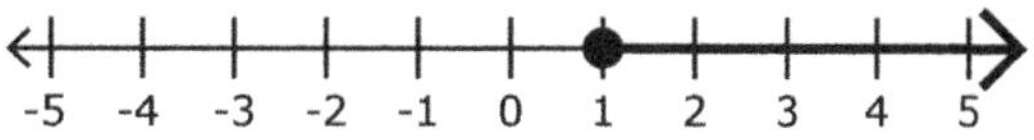

a. $x < 1$
b. $x \leq 1$
c. $x > 1$
d. $x \geq 1$

20. If $-3(x + 4) \geq x + 8$, what is the value of x?
a. $x = 4$
b. $x \geq 2$
c. $x \geq -5$
d. $x \leq -5$

21. What is the solution to the following system of equations?

$$\begin{cases} x^2 + y = 4 \\ 2x + y = 1 \end{cases}$$

a. (−1, 3)
b. (−1, 3), (3, −5)
c. (3, −5)
d. (−1, -5)

22. What is the solution to the following system of equations?

$$x^2 - 2x + y = 8$$

$$x - y = -2$$

a. $(-2, 3)$
b. There is no solution.
c. $(-2, 0)$ $(1, 3)$
d. $(-2, 0)$ $(3, 5)$

23. What is the 42nd item in the pattern: ▲○○□▲○○□▲...?
a. ○
b. ▲
c. □
d. None of the above

24. What is the next number in the following series: $1, 3, 6, 10, 15, 21, \ldots$?

25. Which equation is not a function?
a. $y = |x|$
b. $y = \sqrt{x}$
c. $x = 3$
d. $y = 4$

26. What is the domain for the function $y = \sqrt{x}$?
a. All real numbers
b. $x \geq 0$
c. $x > 0$
d. $y \geq 0$

27. The phone bill is calculated each month using the equation $c = 50g + 75$. The cost of the phone bill per month is represented by c, and g represents the gigabytes of data used that month. What is the value and interpretation of the slope of this equation?
a. 75 dollars per day
b. 75 gigabytes per day
c. 50 dollars per day
d. 50 dollars per gigabyte

28. A ball is thrown from the top of a high hill, so that the height of the ball as a function of time is $h(t) = -16t^2 + 4t + 6$, in feet. What is the maximum height of the ball in feet?
a. 6
b. 6.25
c. 6.5
d. 6.75

29. A line passes through the point (1, 2) and crosses the *y*-axis at *y* = 1. Which of the following is an equation for this line?
a. $y = 2x$
b. $y = 3x$
c. $x + y = 1$
d. $y = x + 1$

30. What is the slope of this line?

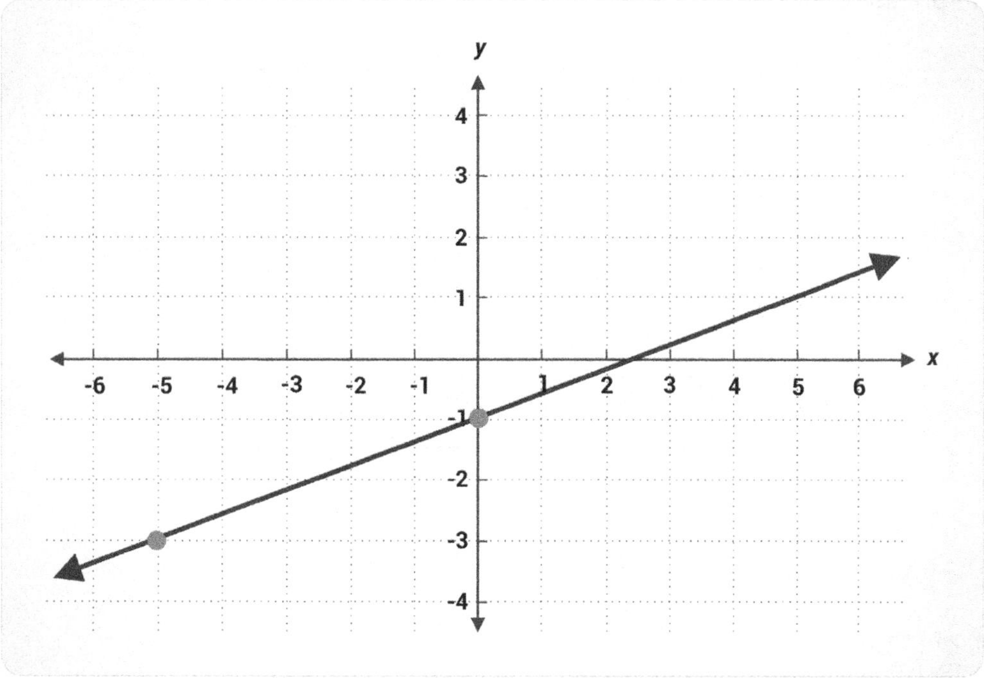

a. 2

b. $\frac{5}{2}$

c. $\frac{1}{2}$

d. $\frac{2}{5}$

31. The graph shows the position of a car over a 10-second time interval. Which of the following is the correct interpretation of the graph for the interval 1 to 3 seconds?

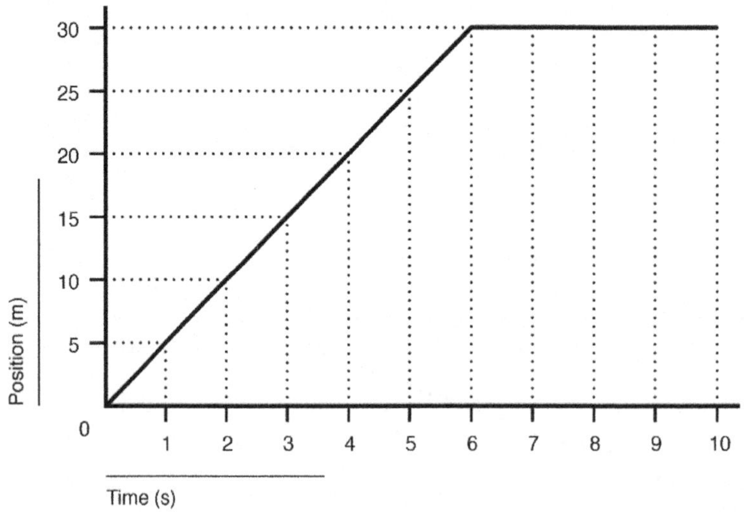

a. The car remains in the same position.
b. The car is traveling at a speed of 5 m/s.
c. The car is traveling up a hill.
d. The car is traveling at 5 mph.

32. The graph of which function has an *x*-intercept of -2?
a. $y = 2x - 3$
b. $y = 4x + 2$
c. $y = x^2 + 5x + 6$
d. $y = -\frac{1}{2} \times 2^x$

33. What is the y-intercept for $y = x^2 + 3x - 4$?
a. $y = 1$
b. $y = -4$
c. $y = 3$
d. $y = 4$

34.

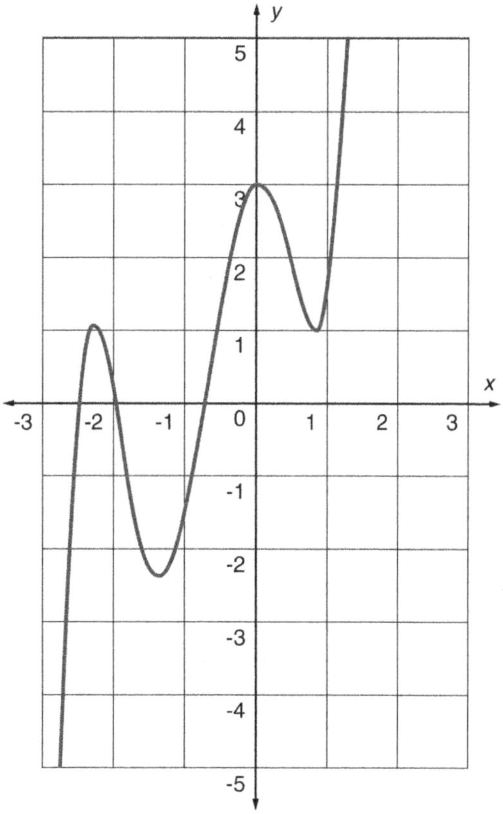

Which of the following functions represents the graph above?

a. $y = x^5 + 3.5x^4 - 6.5x^2 + 0.5x + 3$
b. $y = x^5 - 3.5x^4 + 6.5x^2 - 0.5x - 3$
c. $y = 5x^4 - 6.5x^2 + 0.5x + 3$
d. $y = -5x^4 - 6.5x^2 + 0.5x + 3$

35. What type of function is modeled by the values in the following table?

X	f(x)
1	2
2	4
3	8
4	16
5	32

a. Linear
b. Exponential
c. Quadratic
d. Absolute Value

Geometry and Data

1. What is the volume of a pyramid, with a square base whose side is 6 inches, and the height is 9 inches?
   a. 324 in^3
   b. 72 in^3
   c. 108 in^3
   d. 18 in^3

2. What is the volume of a cone, in terms of π, with a radius of 10 centimeters and height of 12 centimeters?
   a. 400 cm^3
   b. 200 cm^3
   c. 120 cm^3
   d. 140 cm^3

3. If the volume of a sphere is 288π cubic meters, what are the radius and surface area of the same sphere?
   a. Radius 6 meters and surface area 144π square meters
   b. Radius 36 meters and surface area 144π square meters
   c. Radius 6 meters and surface area 12π square meters
   d. Radius 36 meters and surface area 12π square meters

4. For the following similar triangles, what are the values of x and y (rounded to one decimal place)?

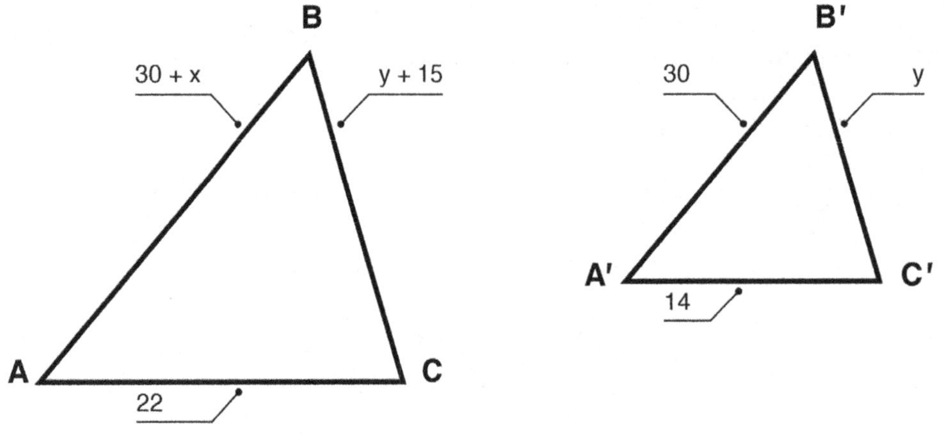

   a. $x = 16.5, y = 25.1$
   b. $x = 19.5, y = 24.1$
   c. $x = 17.1, y = 26.3$
   d. $x = 26.3, y = 17.1$

5. What is the measurement of angle f in the following picture? Assume the lines are parallel.

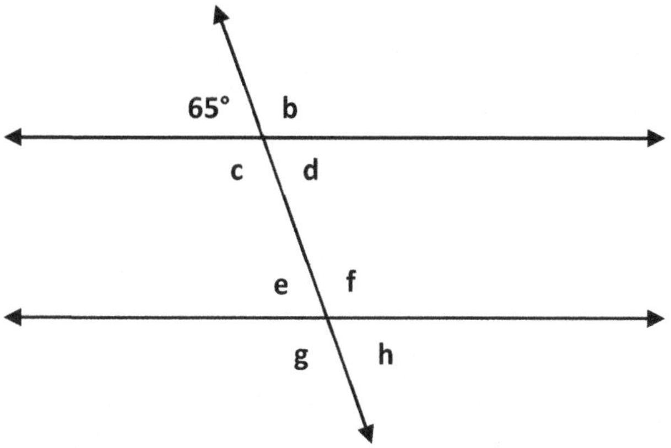

a. 65 degrees
b. 115 degrees
c. 125 degrees
d. 55 degrees

6. The triangle shown below is a right triangle. What's the value of x? Enter your answer in the box below.

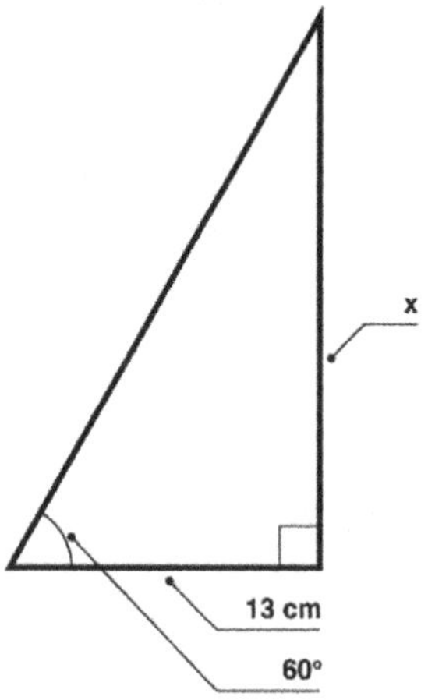

7. An equilateral triangle has a perimeter of 18 feet. If a square whose sides have the same length as one side of the triangle is built, what will be the area of the square?

a. 6 square feet
b. 36 square feet
c. 256 square feet
d. 1000 square feet

8. The area of a given rectangle is 24 square centimeters. If the measure of each side is multiplied by 3, what is the area of the new figure? Enter your answer in the box below.

9. What is the area of the regular hexagon shown below?

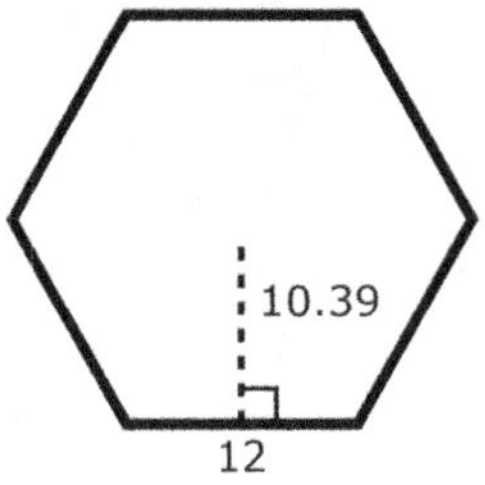

a. 72
b. 124.68
c. 374.04
d. 748.08

10. The square and circle have the same center. The circle has a radius of *r*. What is the area of the shaded region?

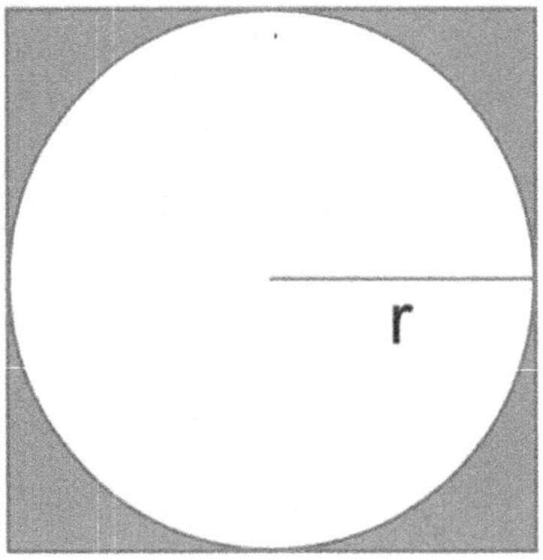

a. $r^2 - \pi r^2$
b. $4r^2 - 2\pi r$
c. $(4 - \pi)r^2$
d. $(\pi - 1)r^2$

11.

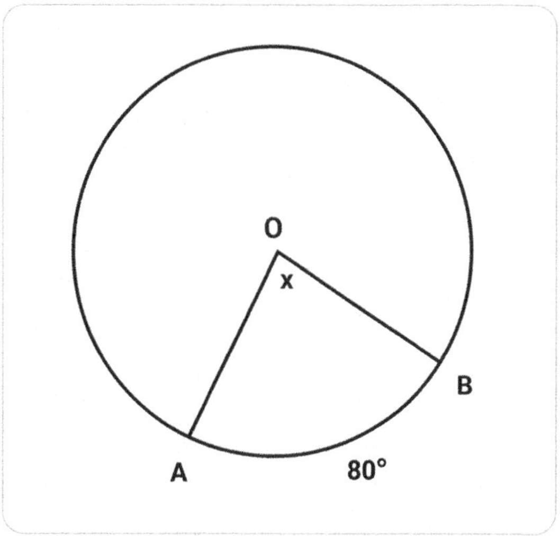

The area of circle O is 49π m. What is the area of the sector formed by $\angle AOB$?

a. 80π m
b. 10.9π m
c. 4.9π m
d. 10π m

12. What's the midpoint of a line segment with endpoints $(-1, 2)$ and $(3, -6)$?

a. $(1, 2)$
b. $(1, 0)$
c. $(-1, 2)$
d. $(1, -2)$

13. If the ordered pair $(-3, -4)$ is reflected over the *x*-axis, what's the new ordered pair?

a. $(-3, -4)$
b. $(3, -4)$
c. $(3, 4)$
d. $(-3, 4)$

14. Mom's car drove 72 miles in 90 minutes. How fast did she drive in feet per second?

a. 0.8 feet per second
b. 48.9 feet per second
c. 0.009 feet per second
d. 70.4 feet per second

15. How many kilometers is 4382 feet?

a. 1.336 kilometers
b. 14,376 kilometers
c. 1.437 kilometers
d. 13,336 kilometers

16. The following graph compares the various test scores of the top three students in each of these teacher's classes. Based on the graph, which teacher's students had the smallest range of test scores?

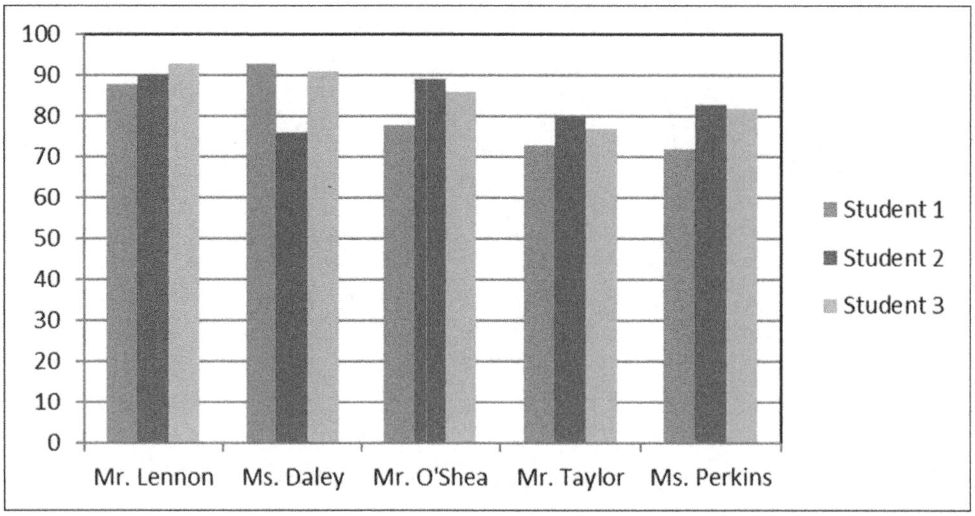

a. Mr. Lennon
b. Mr. O'Shea
c. Mr. Taylor
d. Ms. Daley

17. The chart below shows the percentage (rounded to the nearest tenth of a percent) of seniors from Heights High School and their plans for after high school. The number of seniors planning to attend a four-year college or university is 267. How many total seniors are planning to go into the military? Round your answer to the nearest integer.

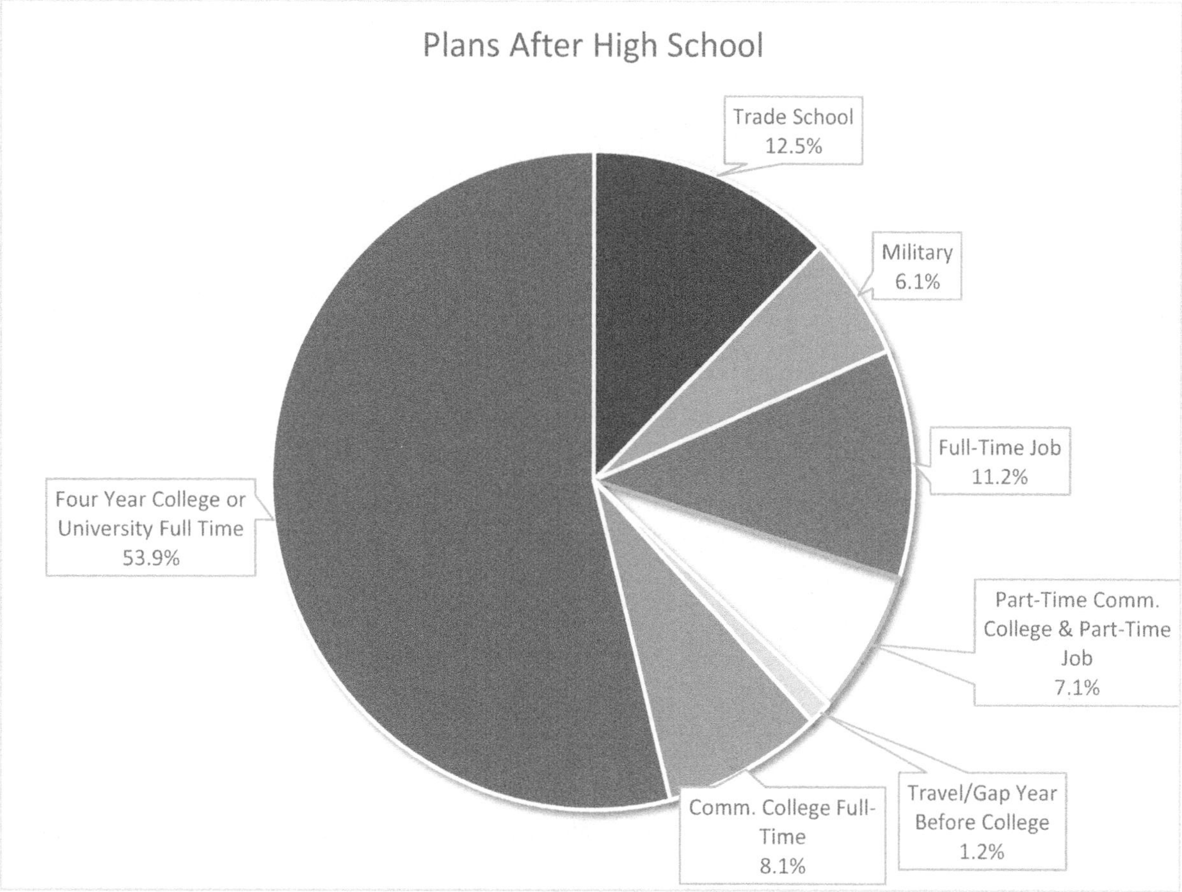

18. The graph below is a scatter plot with 30 points that correspond to the number of miles driven in a given time period for a truck driver this year. The time is given in hours. The dashed line refers to the regression line. Based on the regression line, if the truck driver drives for 7 hours, approximately how far will he go?

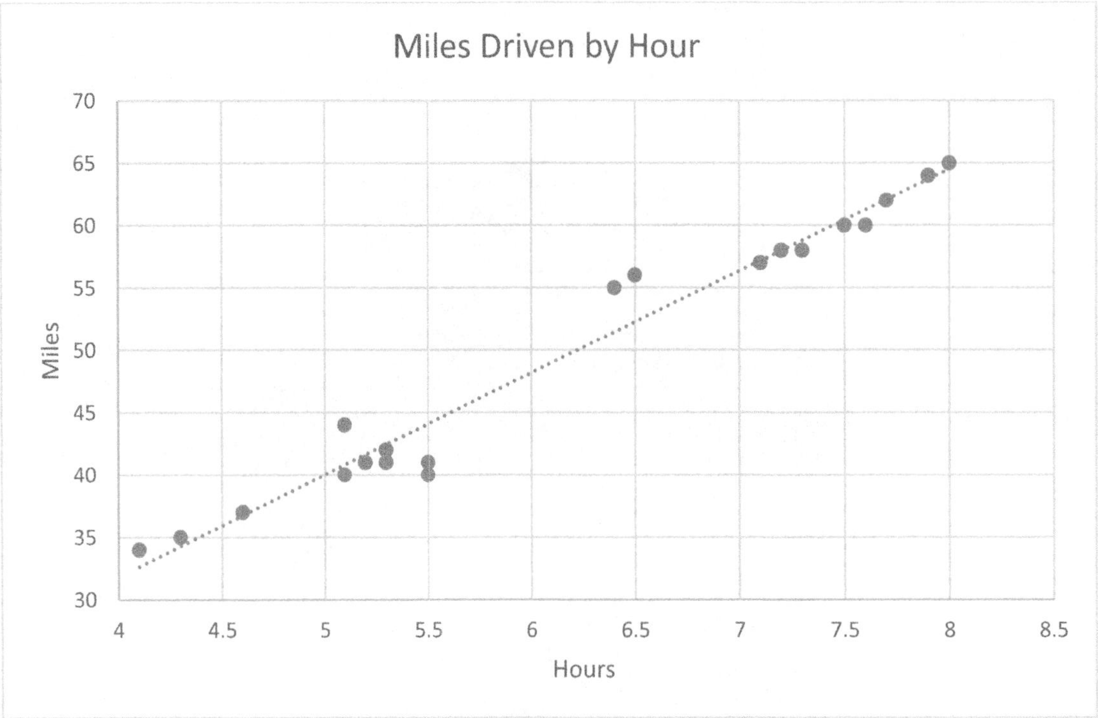

19. Given the value of a given stock at monthly intervals, which graph should be used to best represent the trend of the stock?
   a. Box plot
   b. Line plot
   c. Line graph
   d. Circle graph

20. How many possible two-number combinations are there for the numbers 1, 2, 3, 4, and 5 if each number can only be used once in any combination and order DOES matter?
   a. 120
   b. 60
   c. 20
   d. 10

21. Two cards are drawn from a shuffled deck of 52 cards. What's the probability that both cards are Kings if the first card isn't replaced after it's drawn?

a. $\frac{1}{169}$
b. $\frac{1}{221}$
c. $\frac{1}{13}$
d. $\frac{4}{13}$

22. Ten students take a test. Five students get a 50. Four students get a 70. If the average score is 55, what was the last student's score?

a. 20
b. 40
c. 50
d. 60

23. For a group of 20 men, the median weight is 180 pounds and the range is 30 pounds. If each man gains 10 pounds, which of the following would be true?

a. The median weight will increase, and the range will remain the same.
b. The median weight and range will both remain the same.
c. The median weight will stay the same, and the range will increase.
d. The median weight and range will both increase.

Answer Explanations

Arithmetic and Algebra

1. C: 34. When performing calculations consisting of more than one operation, the order of operations should be followed: *P*arenthesis, *E*xponents, *M*ultiplication/*D*ivision, *A*ddition/*S*ubtraction. Parenthesis: $7^2 - 3 \times (4 + 2) + 15 \div 5 = 7^2 - 3 \times (6) + 15 \div 5$. Exponents: $7^2 - 3 \times 6 + 15 \div 5 = 49 - 3 \times 6 + 15 \div 5$. Multiplication/Division (from left to right): $49 - 3 \times 6 + 15 \div 5 = 49 - 18 + 3$. Addition/Subtraction (from left to right): $49 - 18 + 3 = 34$.

2. 15: Follow the *order of operations* in order to solve this problem. Solve the parentheses first, and then follow the remainder as usual.

$$(6 \times 4) - 9$$

This equals 24 – 9 or 15.

3. B: Scientific notation takes the form $a \times 10^n$, where a is any decimal number between 1 and 10 and n is the number of decimal places it takes to reach the first digit. In this case, take the first digit 7, add a decimal point, and then include each non-zero digit after that to determine the value of a as 7.35. Next, count the place values from the position of the decimal point in standard form to the position in scientific notation in order to determine the value of n, which would be 9. Therefore, this number expressed in scientific notation would be 7.35×10^9.

4. C: The first step in solving this problem is expressing the result in fraction form. Separate this problem first by solving the division operation of the last two fractions. When dividing one fraction by another, invert or flip the second fraction and then multiply the numerator and denominator.

$$\frac{7}{10} \times \frac{2}{1} = \frac{14}{10}$$

Next, multiply the first fraction with this value:

$$\frac{3}{5} \times \frac{14}{10} = \frac{42}{50}$$

Decimals are expressions of 1 or 100%, so multiply both the numerator and denominator by 2 to get the fraction as an expression of 100.

$$\frac{42}{50} \times \frac{2}{2} = \frac{84}{100}$$

In decimal form, this would be expressed as 0.84.

5. 80 min: To solve the problem, a proportion is written consisting of ratios comparing distance and time. One way to set up the proportion is:

$\frac{3}{48} = \frac{5}{x} \left(\frac{distance}{time} = \frac{distance}{time}\right)$ where x represents the unknown value of time

To solve a proportion, the ratios are cross-multiplied:

$$(3)(x) = (5)(48) \rightarrow 3x = 240$$

The equation is solved by isolating the variable, or dividing by 3 on both sides, to produce $x = 80$.

6. B: To be directly proportional means that $y = mx$. If x is changed from 5 to 20, the value of x is multiplied by 4. Applying the same rule to the y-value, also multiply the value of y by 4. Therefore, $y =$ 12.

7. B: A counterexample is a specific example that shows the statement is not true. 2 is prime and an even number.

8. A: The relative error can be found by finding the absolute error and making it a percent of the true value. The absolute error is $36 - 35.75 = 0.25$. This error is then divided by 36—the true value—and then multiplied by 100 (since we are finding the percentage) to find 0.7%.

9. 32: To solve for x, isolate the variables from the constants. First, subtract 13 from both sides:

$$13 - 13 - \frac{3x}{4} = -11 - 13$$

$$-\frac{3x}{4} = -24$$

Next, multiply both sides by 4 to eliminate the fraction:

$$-\frac{3x}{4} \times 4 = -24 \times 4$$

$$-3x = -96$$

Finally, multiply by -3 to solve for x:

$$\frac{-3x}{-3} = \frac{-96}{-3}$$

$$x = 32$$

Therefore, the value of x is 32.

10. C: Kimberley worked 4.5 hours at the rate of \$10/h and 1 hour at the rate of \$12/h. The problem states that her pay is rounded to the nearest hour, so the 4.5 hours would round up to 5 hours at the rate of \$10/h.

$$(5h)(\$10/h) + (1h)(\$12/h) = \$50 + \$12 = \$62$$

11. D:

$9x + x - 7 = 16 + 2x$	Combine $9x$ and x.
$10x - 7 = 16 + 2x$	
$10x - 7 + 7 = 16 + 2x + 7$	Add 7 to both sides to remove (-7).
$10x = 23 + 2x$	
$10x - 2x = 23 + 2x - 2x$	Subtract 2x from both sides to move it to the other side of the equation.
$8x = 23$	
$\frac{8x}{8} = \frac{23}{8}$	Divide by 8 to get x by itself.
$x = \frac{23}{8}$	

12. B: First, calculate the difference between the larger value and the smaller value.

$$378 - 252 = 126$$

To calculate this difference as a percentage of the original value, and thus calculate the percentage *increase*, divide 126 by 252, then multiply by 100 to reach the percentage 50%, Choice *B*.

13. 2: Add 3 to both sides to get $4x = 8$. Then divide both sides by 4 to get $x = 2$.

14. B: $3x^2 - 3x + 11$. By distributing the implied one in front of the first set of parentheses and the -1 in front of the second set of parentheses, the parenthesis can be eliminated: $1(5x^2 - 3x + 4) - 1(2x^2 - 7) = 5x^2 - 3x + 4 - 2x^2 + 7$. Next, like terms (same variables with same exponents) are combined by adding the coefficients and keeping the variables and their powers the same: $5x^2 - 3x + 4 - 2x^2 + 7 = 3x^2 - 3x + 11$.

15. B: To simplify the given equation, the first step is to make all exponents positive by moving them to the opposite place in the fraction. This expression becomes $\frac{4b^3b^2}{a^1a^4} \times \frac{3a}{b}$. Then the rules for exponents can be used to simplify. Multiplying the same bases means the exponents can be added. Dividing the same bases means the exponents are subtracted. Thus, after multiplying the exponents in the first fraction the equation becomes $\frac{4b^5}{a^5} \times \frac{3a}{b}$. Then, the two fractions can be multiplied, and the exponents must be divided to find $\frac{12b^4}{a^4}$.

16. C: Because the two amounts add up to 40, each amount can be found by subtracting the other from 40. If the larger amount is m, the smaller amount is $40 - m$.

17. C: $y = 40x + 300$. In this scenario, the variables are the number of sales and Karen's weekly pay. The weekly pay depends on the number of sales. Therefore, weekly pay is the dependent variable (y), and the number of sales is the independent variable (x). Each pair of values from the table can be written as an ordered pair (*x*, *y*):

(2,380), (7,580), (4,460), (8,620)

The ordered pairs can be substituted into the equations to see which creates true statements (both sides equal) for each pair. Even if one ordered pair produces equal values for a given equation, the other three ordered pairs must be checked. The only equation which is true for all four ordered pairs is $y = 40x + 300$:

$$380 = 40(2) + 300 \to 380 = 380$$

$$580 = 40(7) + 300 \to 580 = 580$$

$$460 = 40(4) + 300 \to 460 = 460$$

$$620 = 40(8) + 300 \to 620 = 620$$

18. A: Parallel lines have the same slope. The slope of *C* can be seen to be 1/3 by dividing both sides by 3. The others are in standard form $Ax + By = C$, for which the slope is given by $\frac{-A}{B}$. The slope of A is 3; the slope of B is 4. The slope of *D* is 1.

19. D: $x \geq 1$. The closed dot on one indicates that the value is included in the set. The arrow pointing right indicates that numbers greater than one (numbers get larger to the right) are included in the set. Therefore, the set includes numbers greater than or equal to one, which can be written as $x \geq 1$.

20. D: $x \leq -5$. When solving a linear equation or inequality:

Distribution is performed if necessary: $-3(x + 4) \to -3x - 12 \geq x + 8$. This means that any like terms on the same side of the equation/inequality are combined.

The equation/inequality is manipulated to get the variable on one side. In this case, subtracting *x* from both sides produces $-4x - 12 \geq 8$.

The variable is isolated using inverse operations to undo addition/subtraction. Adding 12 to both sides produces $-4x \geq 20$.

The variable is isolated using inverse operations to undo multiplication/division. Remember if dividing by a negative number, the relationship of the inequality reverses, so the sign is flipped. In this case, dividing by -4 on both sides produces $x \leq -5$.

21. B: The system can be solved using substitution. Solve the second equation for *y*, resulting in $y = 1 - 2x$. Plugging this into the first equation results in the quadratic equation $x^2 - 2x + 1 = 4$. In standard form, this equation is equivalent to $x^2 - 2x - 3 = 0$ and in factored form is $(x - 3)(x + 1) = 0$. Its solutions are $x = 3$ and $x = -1$. Plugging these values into the second equation results in $y = -5$ and $y = 3$, respectively. Therefore, the solutions are the ordered pairs $(-1, 3)$ and $(3, -5)$.

22. D: This system of equations involves one quadratic function and one linear function, as seen from the degree of each equation. One way to solve this is through substitution. Solving for y in the second equation yields $y = x + 2$. Plugging this equation in for the y of the quadratic equation yields $x^2 - 2x + x + 2 = 8$. Simplifying the equation, it becomes $x^2 - x + 2 = 8$. Setting this equal to zero and factoring, it becomes $x^2 - x - 6 = 0 = (x - 3)(x + 2)$. Solving these two factors for x gives the zeros $x = 3, -2$. To find the y-value for the point, each number can be plugged in to either original equation. Solving each one for y yields the points $(3, 5)$ and $(-2, 0)$.

23. A: ○. The core of the pattern consists of 4 items: ▲○○□. Therefore, the core repeats in multiples of 4, with the pattern starting over on the next step. The closest multiple of 4 to 42 is 40. Step 40 is the end of the core (□), so step 41 will start the core over (▲), and step 42 is ○.

24. 28: Each number in the sequence is adding one more than the difference between the previous two. For example, $10 - 6 = 4, 4 + 1 = 5$. Therefore, the next number after 10 is $10 + 5 = 15$. Going forward, $21 - 15 = 6, 6 + 1 = 7$. The next number is $21 + 7 = 28$. Therefore, the difference between numbers is the set of whole numbers starting at 2: 2, 3, 4, 5, 6, 7….

25. C: The equation $x = 3$ is not a function because it does not pass the vertical line test. This test is made from the definition of a function, where each x-value must be mapped to one and only one y-value. This equation is a vertical line, so the x-value of 3 is mapped with an infinite number of y-values.

26. B: The domain is all possible input values, or x-values. For this equation, the domain is every number greater than or equal to zero. There are no negative numbers in the domain because taking the square root of a negative number results in an imaginary number.

27. D: The slope from this equation is 50, and it is interpreted as the cost per gigabyte used. Since the g-value represents number of gigabytes and the equation is set equal to the cost in dollars, the slope relates these two values. For every gigabyte used on the phone, the bill goes up 50 dollars.

28. B: The independent variable's coordinate at the vertex of a parabola (which is the highest point, when the coefficient of the squared independent variable is negative) is given by:

$$x = -\frac{b}{2a}$$

Substitute and solve for *x* to get:

$$x = -\frac{4}{2(-16)} = \frac{1}{8}$$

Using this value of *x*, the maximum height of the ball (*y*), can be calculated. Substituting *x* into the equation yields:

$$h(t) = -16(\frac{1}{8})^2 + 4(\frac{1}{8}) + 6 = 6.25$$

29. D: From the slope-intercept form, $y = mx + b$, it is known that *b* is the *y*-intercept, which is 1. Compute the slope as $\frac{2-1}{1-0} = 1$, so the equation should be $y = x + 1$.

30. D: The slope is given by the change in *y* divided by the change in *x*. Specifically, it's:

$$slope = \frac{y_2 - y_1}{x_2 - x_1}$$

The first point is (-5, -3), and the second point is (0, -1). Work from left to right when identifying coordinates. Thus the point on the left is point 1 (-5, -3), and the point on the right is point 2 (0, -1).

Now we need to just plug those numbers into the equation:

$$slope = \frac{-1-(-3)}{0-(-5)}$$

It can be simplified to:

$$slope = \frac{-1+3}{0+5}$$

$$slope = \frac{2}{5}$$

31. B: The car is traveling at a speed of five meters per second. On the interval from one to three seconds, the position changes by ten meters. By making this change in position over time into a rate, the speed becomes ten meters in two seconds or five meters in one second.

32. C: An x-intercept is the point where the graph crosses the x-axis. At this point, the value of y is 0. To determine if an equation has an x-intercept of -2, substitute -2 for x, and calculate the value of y. If the value of -2 for x corresponds with a y-value of 0, then the equation has an x-intercept of -2. The only answer choice that produces this result is Choice *C*:

$$0 = (-2)^2 + 5(-2) + 6$$

33. B: The y-intercept of an equation is found where the x -value is zero. Plugging zero into the equation for x, the first two terms cancel out, leaving -4.

34. A: The graph contains four turning points (where the curve changes from rising to falling or vice versa). This indicates that the degree of the function (highest exponent for the variable) is 5, eliminating Choices *C* and *D*. The y-intercepts of the functions can be determined by substituting 0 for x and finding the value of y. The function for Choice *A* has a y-intercept of 3, and the function for Choice *B* has a y-intercept of -3. Therefore, Choice *B* is eliminated.

35. B: The table shows values that are increasing exponentially. The differences between the inputs are the same, while the differences in the outputs are changing by a factor of 2. The values in the table can be modeled by the equation $f(x) = 2^x$.

Geometry and Data

1. C: The volume of a pyramid is ($length \times width \times height$), divided by 3, and ($6 \times 6 \times 9$), divided by 3 is 108 in^3. Choice *A* is incorrect because 324 in^3 is ($length \times width \times height$) without dividing by 3. Choice *B* is incorrect because 6 is used for height instead of 9 (($6 \times 6 \times 6$) divided by 3) to get 72 in^3. Choice *D* is incorrect because 18 in^3 is (6×9), divided by 3 and leaving out a 6.

2. A: The volume of a cone is ($\pi r^2 h$), divided by 3, and ($\pi \times 10^2 \times 12$), divided by 3 is 400 cm^3. Choice *B* is $10^2 \times 2$. Choice *C* is incorrect because it is 10×12. Choice *D* is also incorrect because that is $10^2 + 40$.

3. A: Because the volume of the given sphere is 288π cubic meters, this means ${}^4/_3\,\pi r^3 = 288\pi$. This equation is solved for r to obtain a radius of 6 meters. The formula for the surface area of a sphere is $4\pi r^2$, so if $r = 6$ in this formula, the surface area is 144π square meters.

4. C: Because the triangles are similar, the lengths of the corresponding sides are proportional. Therefore:

$$\frac{30+x}{30} = \frac{22}{14} = \frac{y+15}{y}$$

This results in the equation $14(30+x) = 22 \cdot 30$ which, when solved, gives $x = 17.1$. The proportion also results in the equation $14(y+15) = 22y$ which, when solved, gives $y = 26.3$.

5. B: Because the 65-degree angle and angle *b* sum to 180 degrees, the measurement of angle *b* is 115 degrees. Because of corresponding angles, angle *b* is equal to angle *f*. Therefore, angle *f* measures 115 degrees.

6. 22.49: SOHCAHTOA is used to find the missing side length. Because the angle and adjacent side are known, $\tan 60 = \frac{x}{13}$. Making sure to evaluate tangent with an argument in degrees, this equation gives:

$$x = 13 \tan 60 = 13 \times \sqrt{3} = 22.52$$

7. B: An equilateral triangle has three sides of equal length, so if the total perimeter is 18 feet, each side must be 6 feet long. A square with sides of 6 feet will have an area of $6^2 = 36$ square feet.

8. 216 square centimeters: Because area is a two-dimensional measurement, the dimensions are multiplied by a scale that is squared to determine the scale of the corresponding areas. The dimensions of the rectangle are multiplied by a scale of 3. Therefore, the area is multiplied by a scale of 3^2 (which is equal to 9):

$$24\ cm^2 \times 9 = 216\ cm^2$$

9. C: 374.04. The formula for finding the area of a regular polygon is $A = \frac{1}{2} \times a \times P$ where a is the length of the apothem (from the center to any side at a right angle), and P is the perimeter of the figure. The apothem a is given as 10.39, and the perimeter can be found by multiplying the length of one side by the number of sides (since the polygon is regular): $P = 12 \times 6 = 72$. To find the area, substitute the values for a and P into the formula: $A = \frac{1}{2} \times a \times P = \frac{1}{2} \times 10.39 \times 72 = 374.04$.

10. C: The area of the shaded region is the area of the square, minus the area of the circle. The area of the circle will be πr^2. The side of the square will be $2r$, so the area of the square will be $4r^2$. Therefore, the difference is $4r^2 - \pi r^2 = (4 - \pi)r^2$

11. B: Given the area of the circle, the radius can be found using the formula $A = \pi r^2$. In this case, $49\pi = \pi r^2$, which yields $r = 7$ m. A central angle is equal to the degree measure of the arc it inscribes; therefore, $\angle x = 80°$. The area of a sector can be found using the formula $A = \frac{\theta}{360^o} \times \pi r^2$. In this case, $A = \frac{80^o}{360^o} \times \pi(7)r^2 = 10.9\pi$ m.

12. D: The midpoint formula should be used.

$$M = \left(\frac{x_1 + x_2}{2}, \frac{y_1 + y_2}{2}\right) = \left(\frac{-1+3}{2}, \frac{2+(-6)}{2}\right) = (1, -2)$$

13. D: When an ordered pair is reflected over an axis, the sign of one of the coordinates must change. When it's reflected over the x-axis, the sign of the y-coordinate must change. The x-value remains the same. Therefore, the new ordered pair is $(-3, 4)$.

14. D: This problem can be solved by using unit conversion. The initial units are miles per minute. The final units need to be feet per second. Converting miles to feet uses the equivalence statement 1 mile = 5,280 feet. Converting minutes to seconds uses the equivalence statement 1 minute = 60 seconds. Setting up the ratios to convert the units is shown in the following equation $\frac{72\ miles}{90\ minutes} * \frac{1\ minute}{60\ seconds} * \frac{5280\ feet}{1\ mile} = 70.4$ feet per second. The initial units cancel out, and the new units are left.

15. A: The conversion can be obtained by setting up and solving the following equation:

$$4382\ ft \times \frac{.3048\ m}{1\ ft} \times \frac{1\ km}{1000\ m} = 1.336\ km$$

16. A: To calculate the range in a set of data, subtract the lowest value from the highest value. In this graph, the range of Mr. Lennon's students is 5, which can be seen physically in the graph as having the smallest difference between the highest value and the lowest value compared with the other teachers.

17. 30: Because there are 267 seniors that are planning on going to school full time next year and that equals 53.9% of the total senior class, there are $267 \div 0.539 = 495$ total seniors at Heights High, rounded to the nearest integer. That means that 6.1% of 495 seniors are headed into the military, which equals $495 \times 0.061 = 30$, rounded.

18. 57 miles: The regression line predicts the output, given a specific input. Therefore, to approximate the miles driven in 7 hours, locate 7 on the x-axis, which corresponds to time. Then, locate the regression line at that x-value and find its corresponding y-value. Its corresponding y-value is approximately 57, which refers to 57 miles.

19. C: The scenario involves data consisting of two variables: month and stock value. Box plots display data consisting of values for one variable. Therefore, a box plot is not an appropriate choice. Both line plots and circle graphs are used to display frequencies within categorical data. Neither can be used for the given scenario. Line graphs display two numerical variables on a coordinate grid and show trends among the variables, so this is the correct choice.

20. C: Because order *does* matter, the total number of permutations needs to be computed. $P(5,2) = \frac{5!}{(5-2)!} = \frac{120}{6} = 20$ represents the number of ways that two objects can be arranged from a set of five.

21. B: For the first card drawn, the probability of a King being pulled is $\frac{4}{52}$. Since this card isn't replaced, if a King is drawn first the probability of a King being drawn second is $\frac{3}{51}$. The probability of a King being drawn in both the first and second draw is the product of the two probabilities: $\frac{4}{52} \times \frac{3}{51} = \frac{12}{2652}$. This fraction, when divided by 12, equals $\frac{1}{221}$.

22. A: Let the unknown score be x. The average will be:

$$\frac{5 \times 50 + 4 \times 70 + x}{10} = \frac{530 + x}{10} = 55$$

Multiply both sides by 10 to get $530 + x = 550$, or $x = 20$.

23. A: If each man gains 10 pounds, every original data point will increase by 10 pounds. Therefore, the man with the original median will still have the median value, but that value will increase by 10. The smallest value and largest value will also increase by 10 and, therefore, the difference between the two won't change. The range does not change in value and, thus, remains the same.

Dear Praxis Middle School Math Test Taker,

We would like to start by thanking you for purchasing this study guide for your Praxis Middle School Math exam. We hope that we exceeded your expectations.

Our goal in creating this study guide was to cover all of the topics that you will see on the test. We also strove to make our practice questions as similar as possible to what you will encounter on test day. With that being said, if you found something that you feel was not up to your standards, please send us an email and let us know.

We would also like to let you know about other books in our catalog that may interest you.

Praxis Elementary Education Test

This can be found on Amazon: amazon.com/dp/162845847X

Praxis English Language Arts

amazon.com/dp/1628454105

Praxis Social Studies

amazon.com/dp/1628457686

Praxis Mathematics

amazon.com/dp/1628454261

Praxis Core Study Guide

amazon.com/dp/1628454946

Praxis General Science Study Guide

amazon.com/dp/1628454385

We have study guides in a wide variety of fields. If the one you are looking for isn't listed above, then try searching for it on Amazon or send us an email.

Thanks Again and Happy Testing!
Product Development Team
info@studyguideteam.com

FREE Test Taking Tips DVD Offer

To help us better serve you, we have developed a Test Taking Tips DVD that we would like to give you for FREE. **This DVD covers world-class test taking tips that you can use to be even more successful when you are taking your test.**

All that we ask is that you email us your feedback about your study guide. Please let us know what you thought about it – whether that is good, bad or indifferent.

To get your **FREE Test Taking Tips DVD**, email freedvd@studyguideteam.com with "FREE DVD" in the subject line and the following information in the body of the email:

a. The title of your study guide.

b. Your product rating on a scale of 1-5, with 5 being the highest rating.

c. Your feedback about the study guide. What did you think of it?

d. Your full name and shipping address to send your free DVD.

If you have any questions or concerns, please don't hesitate to contact us at freedvd@studyguideteam.com.

Thanks again!